What others are saying about Jason
Contractor's Survival Guide

M000103059

"If you read this book, it's very clear Jason Reid knows how to grow and run a business in ways that enhance both the top-line and bottom-line along the shortest time line. The only thing that could possibly be better is to hear Jason telling it live, which I've had the pleasure to do on many occasions. Jason Reid knows his stuff and we would all benefit from a 'good listen'."

Jack Daly, CEO
Professional Sales Coach, Inc.

"Ambassador Energy has been extremely fortunate to have Jason Reid on our Advisory Board. From the beginning of our growth phase, Jason offered steps and tactics which are illustrated in this well-constructed, insightful book. We have experienced the benefits of Jason's intellectual property, and I highly recommend it in the form of his book, as he clearly articulates strategy, reason, motivation and many useful nuances to recognizing success in all markets—not just challenging ones!"

Steve Fulgham
CEO, Ambassador Energy, Inc.

"A fantastic blueprint for creating a long-term sustainable business for any industry. An incredible resource by someone who has a proven track record for utilizing these simple yet extremely successful principles in building his own business empire."

John R. DiJulius III
Author
What's The Secret? To Providing a World Class Customer Experience

"As a CEO, coach and speaker, I read an enormous number of books and I can highly recommend this book. The *Survival Guide* is jam packed with fantastic information to grow your business. The question is not whether the book's ideas will work, but whether you will act on them to make huge strides in your business."

Chris Thomson
CEO, Speaker, Coach

"Thank you, thank you, thank you for writing this book! It covers all of the basic, absolutely essential principles for any contractor to survive in this economy. I cannot tell you how many contractors will benefit from this book. Your book should be required reading for any person looking to start their own business!"

Kari Broy
Executive Accounts Manager
Find Me A Local Painter

"This book will help all contractors, but particularly those trying to grow their businesses. Jason has done a great job in laying out, in an easy to read format, the steps to success for those willing to grow."

Richard Peckham
Chairman of the Board
Westside Building Material Corp.

"The *Contractor's Survival Guide* is a "to the point," at times "fiercely to the point," direct attack on bad business. This book is applicable to many small entrepreneurs in many industries, not just those in construction contracting. If you are a small business owner, if you are looking to start a small business, or if you are in school and you are considering the immense opportunities available to you if you become an entrepreneur—GET THIS BOOK! Jason Reid provides a well thought out and direct insight into the missed opportunities of small business today, then provides a road map to a style that provides greater success and prosperity. The opportunities are endless when the chains of disorganization and small thinking are shed."

Matthew K. Stewart
Chairman of the Global Board
The Entrepreneurs' Organization

"Jason Reid's ability to see through the "contractor tunnel vision" we all have as business owners is unbelievably beneficial to us all. His book is filled with many important "ah-ha" moments we as small businesses need to improve upon. It prepares the everyday mom and pop contractor for best practices both in and out of the field. The insight and ideas definitely have real-world applications, many of which I have implemented in my small solar business. This book is a definite read for any home improvement contractor, large or small, looking to build a profitable, efficient and dynamic business."

Ricky Chu
Director of Operations
Frontier Solar Inc.

"As a student of Mr. Reid's for the last 15 years, I can assure you that no one knows the home improvement business better. He has been paramount in helping me build three businesses in the remodel market and I am thrilled that he has finally taken the time to write this book. His ideas are simple, many of them are timeless, and several of them are on the cutting edge of new technologies; especially his approach to marketing and how to build a solid Internet presence. You can learn all the lessons the hard way, as many of us do, or you can read his book, study it, read it again, and keep it close at hand. Without a doubt, this is the best how-to manual ever written for small business owners."

Matt Landauer
Owner—ProWorks Flooring

"Real stuff for real life. A must read to survive! Thank you."

Daniel G. Martucci
Specialized Services Group

"The *Contractor's Survival Guide* hits the nail on the head, literally. Jason has done a fantastic job of getting right to the core of what is important for home improvement contractors everywhere. A must read for all."

Mark Moses
CEO Coach, Speaker
"Coaching Entrepreneurs to Extraordinary Results"

"In my 25 years in the contracting business, I've never come across a more practical guide to success in our industry. The *Contractor's Survival Guide* is much more than just a survival guide . . . it provides the framework for long-lasting success in the contracting business. The *Contractor's Survival Guide* is full of time-tested, practical guidelines that can give a company (big or small) the competitive advantage to rise above the rest. It's a must have if you're in any contracting business. It's extremely rare to find such a complete collection of practical knowledge about the contracting business in just one book. Implementing just a few of the ideas in the *Contractor's Survival Guide* will save us thousands of dollars."

Randy Nelson
Fire and Security Services

CONTRACTOR'S SURVIVAL GUIDE

Building Your Business in Good Times and Bad

By Jason Reid
Co-Founder, National Services Group

DELMAR
CENGAGE Learning™

Australia • Brazil • Japan • Korea • Mexico • Singapore • Spain • United Kingdom • United States

DELMAR
CENGAGE Learning™

**Contractor's Survival Guide:
Building Your Business in Good
Times and Bad**

Vice President, Technology and
 Trade Professional Business
 Unit: Gregory L. Clayton

Director of Building Trades: Taryn
 Zlatin McKenzie

Acquisitions Editor: Robert Person

Product Manager: Vanessa L. Myers

Director of Marketing: Beth A. Lutz

Marketing Manager: Marissa Maiella

Production Director: Carolyn Miller

Production Manager: Andrew
 Crouth

Content Project Manager: Brooke
 Greenhouse

Art Director: Benjamin Gleeksman

For product information and technology assistance, contact us at
Cengage Learning Customer & Sales Support, 1-800-354-9706

For permission to use material from this text or product,
submit all requests online at **www.cengage.com/permissions**.
Further permissions questions can be e-mailed to
permissionrequest@cengage.com

Library of Congress Control Number: 2010936619

ISBN-13: 978-1-111-13540-9
ISBN-10: 1-111-13540-1

Delmar
5 Maxwell Drive
Clifton Park, NY 12065-2919
USA

Cengage Learning is a leading provider of customized learning solutions with office
locations around the globe, including Singapore, the United Kingdom, Australia,
Mexico, Brazil, and Japan. Locate your local office at: **international.cengage.com/region**

Cengage Learning products are represented in Canada by Nelson Education, Ltd.

Visit us at www.InformationDestination.com

For more learning solutions, please visit our corporate website at www.cengage.com

Notice to the Reader
Publisher does not warrant or guarantee any of the products described herein or
perform any independent analysis in connection with any of the product information
contained herein. Publisher does not assume, and expressly disclaims, any obligation
to obtain and include information other than that provided to it by the manufacturer.
The reader is expressly warned to consider and adopt all safety precautions that
might be indicated by the activities described herein and to avoid all potential
hazards. By following the instructions contained herein, the reader willingly assumes
all risks in connection with such instructions. The publisher makes no representations
or warranties of any kind, including but not limited to, the warranties of fitness for
particular purpose or merchantability, nor are any such representations implied with
respect to the material set forth herein, and the publisher takes no responsibility
with respect to such material. The publisher shall not be liable for any special,
consequential, or exemplary damages resulting, in whole or part, from the readers'
use of, or reliance upon, this material.

Printed in the United States of America.
1 2 3 4 5 6 7 12 11 10

To my business partners, for their support and understanding
as to why writing this book was important to me.

To my wife, Kim, and children, Derek, Ashlyn, Kyle, and Ryan,
for allowing me to take our family time to put this book to paper.

CONTENTS

AUTHOR BIOGRAPHY

Jason Reid is cofounder of National Services Group. National Services Group operates as College Works Painting and as Empire Community Painting. Together they are one of the largest residential repaint contractors in the United States. Jason is also on the board of Ambassador Energy, a fast-growing solar solutions company. Jason and his partners have twice been finalists for the Ernst and Young "Entrepreneur of the Year" awards.

DOING BUSINESS TODAY

THE CONTRACTOR DEATH SPIRAL

It all starts with losing a job you thought you were going to get. The phone is not ringing the way it used to be ringing, and there just doesn't seem to be as much work out there now as there was a month ago or a year ago. So the spiral begins. . . .

- You are running out of work for your guys.
- The next quote has to be booked or there is no work next week. To get the job, you lower your prices.
- You start to believe that you can now only sell jobs at the lower price point; you are afraid to go back to your regular rates.
- After a month or two, you realize that your guys are busy, but you are not making any money.
- To save money, you go to work on the job. At least you are doing what you love, and you're making a little profit.
- The work is more tiring than you remember it being, and you are now a lot more tired at night. There is no time to go and find more work. It is becoming harder and harder to return phone calls and emails because you are on the job site all day.
- You miss a couple of quotes because you did not get back to the customer soon enough.
- There's not a ton of work out there, and you need to make some money.
- You lay off more people and do more of the work yourself to make what work you do have last longer.
- There's no money for that new marketing idea. Oh, well, you're too busy anyway.
- There sure are a lot less headaches with no employees.
- You're not sure what you'll do when this last job is finished.
- You wonder who is hiring.

MOVING FORWARD

Unless you live in a cave in the middle of nowhere, you are all too aware that the world we live in is different today than the world we lived in just last year.

Every day there are new headlines about more foreclosures, the stock market reaching new lows, companies—big and small—laying off more workers, others finding that they can do just fine with a reduced workforce, and various articles about how long we will all be in this financial funk. It is enough to make people grab the kids and any cash they have and move to a place where they can hunt their own food and grow their own vegetables.

The recessionary paralysis that many business owners fall into spreads across almost every industry. Everywhere you look, people seem to be cautious and scared. Every business owner you speak to is cutting back to make sure they can "weather the storm" and everyone seems to be hunkering down for a long winter's night or years of intrepid weather.

Not me. I relish these types of economic environments. During the boom times anyone could make a dollar, most of the time just by showing up. Customers would buy whatever you had, whether you were good at what you did or not, or were just the first one to show up to the party.

The competition is different during boom times. Everyone is starting a new business and everyone is an expert. New businesses pop up everywhere almost every day. When the unemployment rate is below 6%, employees are in the driver's seat. Raises that are ridiculous in size and proportion become the norm as owners try desperately to hold on to anyone who is willing to work. Employee quality and pride of workmanship are dropped as they know they can always go down the street for a $1 more an hour if they don't like working for you.

When times get tight, though, the fly-by-night operators all but disappear. The companies that survive are the ones that focus on quality, customer service, and value for their clients. Work ethic trumps luck, and money can be poured into training of employees because they will appreciate it more and are less likely to jump ship since opportunities are few and far between! These are the times to capitalize on the fact that the bar is very low in the home improvement industry, and most of your competitors are closing their doors or shrinking down to one- and two-man shops. There really is no better time to be in business!

About 90% of the population is not all that affected by a recession. They may have less equity in their homes, but they still own them. They still have jobs that need to be done and things will still need to be repaired, maintained, and upgraded in that great big investment we all call "home, sweet home."

You don't need 97% of the population to run a successful business—you can do just fine with the 90% who are still in the market.

The great thing about hard economic times is that your competition has all but disappeared. If you run a legitimate, well-organized, and professional company, there is even more work out there for you to acquire!

This book will help you make sure you are doing all the things that you need to do to run a first-class organization and grow during both the good times and the bad. Keep this book with you at all times and refer to it daily. If you follow all the steps that are outlined here, you will have a successful business that will be the envy of all your competitors!

> *The great thing about hard economic times is that your competition has all but disappeared.*

Why Small Contractors Stay Small

I have spent the last 20 years learning how to be a contractor and a businessperson. During those 20 years, my partners and I have built our company, National Services Group. We started from nothing, and today we are one of the largest residential repaint contractors in the country with thousands of employees spread across the United States doing approximately $40 million in business per year.

Among the many things I have learned over the years is that the words *contractor* and *businessperson* are not interchangeable.

Contractors are, by nature, some of the hardest-working individuals out there. They put in long hours doing backbreaking work (even though they know they should be using their minds more and their bodies less). They care greatly about the finished product and have a love of their chosen trade. There are some contractors who build $5 million, $25 million, and $100 million businesses, but the vast majority of them stay at under $500,000 in business, spending their lives in what some of us would call the definition of insanity—doing the same thing over and over again and achieving the same result. The *result* is a constant struggle to make payroll, pay the bills, and figure out how to get to the next level.

I have spent my career watching small contractors and talking with them. I have watched the mistakes they make (I made most of them myself at one point or another). I have witnessed their solutions and the strategies they try to implement and have seen the results of those efforts. I have also asked myself why it seems to be so hard for contractors to build the business they have dreamt about most of their adult lives. I believe I have figured out over the years not just the reason for this difficulty, but the solution as well.

> *I have also asked myself why it seems to be so hard for contractors to build the business they have dreamt about most of their adult lives.*

They Love Their Trade More Than They Love Business: Most home improvement contractors end up owning their businesses by going down very similar paths. They ended up making a profession out of a trade that they started to enjoy. They became good at that trade and eventually good enough that they thought that they could run the business better than the guy that they worked for.

Most also had the unfortunate misconception that the people that they were working for made money hand over fist. They quickly discovered, though, as they started out on their own, how erroneous that thinking was.

This Book Is for You Regardless of Your Company Size: When the time comes that you decide to run a business, you need to make some decisions. Do you want to run a business that is a good-paying job where you have yourself and a helper or two? Do you want to run one, two, three, 10, or 20 projects at a time? Are you OK not working *in* your trade anymore and, instead, spending your time using your mind rather than your hands? What is really important to you?

Regardless of how you answer these questions, this book is valuable to you. The principles of running a solid first-class business are still the same whether the business employs two people or 2,000. So if you decide that you want to be extremely hands-on during all of your projects, or that you eventually would rather manage the business by looking at reports generated from the systems you develop, the basic building blocks are all right here in this book.

> *The principles of running a solid first-class business are still the same whether the business employs two people or 2,000.*

How We Ended Up Here

Love of the Trade: You are in the business you are in simply because you enjoy your chosen trade. That love of your trade has benefits with regards to running your business. The most important benefit is that you truly understand everything that is important about the job you are doing. Your love of the trade means that your quality is high and your customers (assuming you follow the customer service rules) are satisfied with you and refer new work to you.

You Fell into It: I never had a love of painting. I started painting in the summer when I was 16 at a pharmaceutical plant in Quebec for which my father worked. I was the only English-speaking person at an all-French-speaking unionized plant. This meant that I was given every job that no one else wanted to do. Hence, I spent most of my summers painting.

I have great stories from those days, but basically they put me in the basement, on the roof, and on ladders painting pipes and anything else that needed to be painted. That was my job. When I did it too fast, I got in trouble for making everyone else look bad. When I took a nap because I was done too fast, I'd get criticized for that as well. You get the picture.

I moved away to go to college and got a job painting houses for the summer. Why? Because I needed money to pay for college and I knew how to paint. I stayed in the painting business not because of a love of the trade, but because it was what I knew how to do and I had a love of business.

When I was younger, I remember hearing people say that, after all, Procter & Gamble® just sells toothpaste and Coca-Cola® just sells flavored sugar water. Business is business. And a service business will always be needed. I decided that painting was a service that almost everyone needed and where the competition was not very tough.

Wanting to Be Your Own Boss: At some point in time, some of you decided that you really wanted to be your own boss. Running a home improvement business was a great way to do that. In fact, a very high percentage of contractors end up in their businesses primarily because they were tired of working for someone else and felt that they could do a better job.

How Low Is the Bar in Your Industry?:   This is where I will offend many of you. What I love about the business I am in, and the contracting business in general, is that the bar has been set so low in this industry. In many cases, all you have to do is call your clients back on time and show up on time and the job is yours.

Consumers have been conditioned over the years to expect the lowest levels of customer service from most contractors.

Consumers have been conditioned over the years to expect the lowest levels of customer service from most contractors. If you are just slightly better than terrible, you're going to win the contract. These customers follow the line of thinking that comes from the saying, "If you give a starving man a cracker, that cracker will be the best darn cracker he ever ate!"

This low bar is one of the great reasons to be working in the home improvement industry. By following the basic concepts in this book, you will be miles ahead of your rather incompetent competition.

You—CEO and Owner

It is lonely at the top. Regardless of why you chose to do what you are doing, you are now where you're at.

You have ultimate responsibility for the success or failure of your business.

You are the one who goes without a paycheck in order to make sure your people get paid.

You are the one who cares the most about each and every customer. You are the one who will be here for the long run.

You are the one who lies in bed at night and cannot sleep as everything from new ideas to stressful thoughts race through your head.

You are the one.

Working On Your Business, Instead of In It: As CEO you have a responsibility to everyone who works in your organization. You need to understand what your job actually is.

At times you are the "chief cook and bottle washer."

At times it may seem like you are the only one who gets it.

At times it may seem like you are the one who needs to do all the work.

The most important job you have, though, is to take the time to step back and work on the business. Working on the business means that you are not working on the day-to-day operations of the business, but are using your mind to think and plan where your business is headed.

Working on the business means spending time to work on new ideas such as marketing strategies, sales training, two-year projections, etc., not always just about getting today's job done. It means finding time to think, read, learn, talk to mentors and advisors, and to figure out just how others built their companies.

Your Dream: You started your company because you had a dream of building a great business. It may have been to build a better company than all of your competitors. It may have been to make a bunch of money or to be partners with your kids some day. Whatever the reason, chances are, as time goes by, things have not exactly ended up as you planned.

Maybe your company has exceeded your expectations. But, more than likely, it has frustrated you because you are not quite where you had always planned on being. Not meeting your expectations tends to be the natural course of business.

A wise man once said that entrepreneurs tend to overestimate what they can do in five years and underestimate what they can do in 15. Staying in the game and working every day to make progress are what running a business is all about.

I hope this book helps you on your journey to success—however you define success.

> *Whatever the reason, chances are, as time goes by, things have not exactly ended up as you planned.*

What This Book Is and Is Not

This book is short and to the point.

This book is heavily weighted toward finding business and growing your company.

This book is fast paced and without fluff.

This book will give you great ideas you can use *now*.

This book will help grow your business as you implement its ideas.

This book is not a magic feather. You still need to do the hard work.

This book is not an encyclopedia of everything about each topic.

This book is not all there is to learn on any topic.

This book will not do the hard work for you.

That is up to you.

CHAPTER 1

BRANDING

Branding, marketing, and lead generation. Really they are all the same, right? No, no, no! One of the most important lessons in generating business for your company is to understand that branding, marketing, and lead generation are all uniquely different and play different, although important, roles in building your business. Throughout the next three chapters you'll learn the basics that you need to understand about each of these unique concepts in order to help you build your business.

BRANDING IS ALL ABOUT WHO YOU ARE AND WHAT YOUR COMPANY STANDS FOR

How do you want to be known? How do you want your customers to remember you, and what do you want your reputation to be? Some examples of branding are:

- **Empire Community Painting**—*The company that's easy to do business with!*™
    Empire's goal is to make the painting experience easy for its customers. Empire wants to communicate to potential customers that it takes all of the typical headaches that are generally associated with painting jobs out of the customer experience.
- **Mike Diamond Plumbing (Los Angeles, CA)**—*"I guarantee my plumbers will smell nice."*™
    Mike Diamond Plumbing wants to make the point that its plumbers are not like other plumbers and that its customers have a much more enjoyable and professional experience.

While these examples are really "tag lines," they are designed to show what the company stands for and how, in just one simple sentence, the potential customer can identify with the company.

How do you want to be known? How do you want your customers to remember you, and what do you want your reputation to be?

Branding Exercise

This exercise will help you decide what your company stands for and ultimately what your brand and tag line will be.

- List three things in which your company excels and why:

    1. _____

    2. _____

    3. _____

- List three things that you want your company to be known for:

    1. _____

    2. _____

    3. _____

- List the five things about your company's services you believe are most important to your customers:

    1. _____

    2. _____

    3. _____

    4. _____

    5. _____

- Now comes the hard part. Go out and visit five of your customers and ask them what they think are the most important aspects of working with someone in your trade. Take a note pad and be ready to write all of them down.

    1. _____

    2. _____

    3. _____

    4. _____

    5. _____

After you have interviewed your customers, take a look at the 25 things that your customers listed as the most important aspects they look for in choosing a business in your trade. Chances are you will see some common themes. You will probably be able to narrow these 25 items down to five common themes.

Now compare these five common themes with what you listed as what you believe that customers think are the most important aspects of dealing with your business. If you are really in tune with your customers, the lists will match.

Next we need to look at the first two lists. In a perfect world, what you want your company to be known for is what your customers told you they want and is also what your company is best at doing. If this is the case, then we are ready to pick a tag line for your company. If you and your customers are not aligned, it is time to go back and retool your company.

One of the many keys to building a successful contracting company is to truly offer a service in a manner that your potential customers want. The key to building a contracting business is to effectively communicate to your customers that you deliver exactly what they seek.

If what your company is best at and what you want your company to be known for are not matching up with what your customers want, then you really do need to do some soul-searching. Are you out of touch with your customers? Are you marketing to the right customers? Is it time to rethink how you run your business?

For example, if you think your company is great at solving some very challenging problems that no one else can solve, but your customers are really more interested in great customer service, then your business and your customers are out of alignment.

Or, if you think one of your greatest strengths is that you are an expert with a specific type of material, but your customers believe that a solid mix of quality and price is most important, then your business and your customers are out of alignment.

In general, the company that has the best branding is the company that truly understands what is important to their potential customers.

> *In general, the company that has the best branding is the company that truly understands what is important to their potential customers.*

YOUR BRAND

Your brand should be what your company stands for and hopefully matches with the top five things that are important to your customers. Your brand will sometimes be defined in an entire paragraph on your Website or on your marketing materials, but should also be boiled down to one sentence that states the most important thing about your company. This is your company tag line.

Tag Line

At this point you have taken the time to really understand what your customers want, and you have taken the steps to make sure that what your customers want actually matches with what you do.

Here comes the creative moment. The tag line is the one sentence that you want your company to be known for. It should be simple, but needs to make this point as a reminder to your customers that you know what is really important to them.

Some simple tag lines are:

- *Built ram tough for the long haul.*™ – Dodge Heavy Duty Diesel Truck Engine®
- *All the news that's fit to print.*™ – The New York Times®
- *It takes a licking and keeps on ticking.*™ – Timex®
- *See what brown can do for you.*™ – UPS®
- *Have it your way.*™ – Burger King®
- *M'm! M'm! Good!*™ – Campbell's Soup®

- *When it absolutely, positively has to be there overnight.*™ – Federal Express®
- *The quicker picker upper.*™ – Bounty®
- *Capitalist Tool.*™ – Forbes®

Write down five simple phrases that could be your tag line:

1. _____
2. _____
3. _____
4. _____
5. _____

Not as Easy as it Sounds?

Then get some help. You will be amazed at how creative the people around you are. Your employees, family, customers, and friends can help. Show them the list of five things that you want to be known for and have a contest. Offer $100 to the one that comes up with the best tag line for your company. You will be amazed at what they contribute.

Offer $100 to the one that comes up with the best tag line for your company.

The Choice is Yours

Once you have a list of five or more options from yourself, friends, employees, and family, choose the one that best fits your own style. It's your decision. After all, this is your company!

Logo

Logos are simply the look of your company name. Logos are meant to be something that is recognizable and memorable. The important thing to remember here is that logos can mean an awful lot if you are, or want to be, a recognizable national brand. We all recognize the giant "M" of McDonald's®, the Disney® logo, etc. The reality is that, for the most part, none of us are creating national brands that need the perfect logo.

There are companies that specialize in logo creation. Most of these companies can give you inexpensive options that will include your company name and tag line in a way that will help you communicate who you are and make your company look professional and reliable.

SUCCESSFUL BRANDING

The key to successful branding is to make sure that from this day forward everything that leaves your office with your company information on it has the exact same look and feel. Your logo, tag line, and who you are should never change. Even in a small market, you want your customers to easily recognize you and not be confused by a mess of different marketing materials, logos, or tag lines. All of your marketing materials, signage, Websites, invoices, etc. should have the same look and feel. Make a commitment to this new image of yours and stick to it.

Your logo, tag line, and who you are should never change.

BRANDING—THE SUMMARY

We have now defined who your company is and what it stands for. At this point you should have:

- **A defined brand:** These are the three to five points that summarize who you are as a company. These were created by matching up what you are best at with what is most important to your customers.
- **A tag line:** This is a single sentence that summarizes who you are for your customers.
- **Your logo:** Your company name is presented in a clean and professional graphic that incorporates your tag line.
- **The same all the time:** Keep your branding the same all of the time. People will eventually start to remember your company's look and tag line, even in a small market.

YOUR ELEVATOR PITCH

Yes, this is the overused phrase from the dot-com days where you were to be able to pitch your company successfully to a venture capitalist by the time the elevator hit the 30th floor. Those dot-com days are gone, but the concept remains. You and all of your employees need to be able to sum up your company and what it stands for in three or four succinct sentences.

You are essentially taking your new "brand identity" and vocalizing it. This will not necessarily feel very comfortable in the beginning for either you or your employees. Practice makes perfect.

You and all of your employees need to be able to sum up your company and what it stands for in three or four succinct sentences.

CHAPTER 2

MARKETING

The challenge in defining what marketing is has nothing to do with its definition.

Marketing is defined by the American Marketing Association as *the activity, set of institutions, and processes for creating, communicating, delivering, and exchanging offerings that have value for customers, clients, partners, and society at large.*

You knew that already, right?

The confusion comes from what most marketing companies and "gurus" say you need to do in order to get new business.

Many marketing gurus will tell you that a consumer needs to see your company's name in front of them something like 29 times before they are comfortable enough to use you. This may very well be true if your goal is to build a national brand—or even a local or regional brand where you are trying to make your company name synonymous with the sector you are in.

The reality is that this strategy may be great for Coca-Cola®, Pepsi®, Taco Bell®, Johnson & Johnson®, etc., but it is a sure way to bankrupt a home improvement contractor.

Marketing for the rest of us can be described as simply getting your name out to potential customers and communicating to them what it is that you offer and how you can help them. Following the "KISS" theory (Keep It Simple, Stupid), here are the marketing basics that will help move your company up to the next level.

> *Marketing for the rest of us can be described as simply getting your name out to potential customers and communicating to them what it is that you offer and how you can help them.*

A TALE OF THREE STRATEGIES

Contractors fall into three camps:

1. Those who believe they have a "word-of-mouth" business and do nothing at all in terms of marketing.

2. Those who spend money on marketing and are constantly disappointed with the results.

3. Those who have read this book.

STRATEGY #1—DOING NOTHING AND LOVING IT

It is possible to keep yourself busy by doing high quality work and relying on the kindness of your customers to refer you to their neighbors. This strategy may keep you (and maybe one other worker) busy during times of economic prosperity, but it won't grow your business.

Using this "word-of-mouth" approach helped contractors 40 years ago because people actually spoke to their neighbors. Today people are becoming more and more isolated and family focused. It is not uncommon for neighbors to live in a suburban neighborhood for five or 10 years and never really know their neighbors. They may share an occasional wave as they pass each other pulling in or out of the driveway as they quickly slide in or out of the garage, but rarely do they spend any time together.

Consequently, you may have just done a terrific job for somebody, but if that customer doesn't talk to his neighbors, the hoped-for endorsement is a best-kept secret. There is, unfortunately, an exception to that rule. Nothing brings a neighborhood together like a horror story of how bad of a job a contractor just did. Neighbors who have not spoken two words to each other in six years suddenly are best friends, bonded around the story of how you destroyed their home. (More about that later.)

Word-of-mouth contractors are also the first contractors to suffer during a downturn in the economy. It only makes sense. If less people are getting work done, then there are less people to talk about you.

During down times, people tend to want to protect friends and neighbors and will not brag about how great things are at their home and how much they enjoy their latest home improvement project. They do not want to come across as gloating to someone who is not doing as well as they are. Or worse, the person they just told about their new addition might decide they're a good source of cash to borrow from!

In reality, during tough times, "word-of-mouth" contractors are hit first and hit the hardest. Those contractors who are doing no marketing and "loving what they do" generally end up doing nothing at all.

> *Using this "word-of-mouth" approach helped contractors 40 years ago because people actually spoke to their neighbors.*

> *Neighbors who have not spoken two words to each other in six years suddenly are best friends, bonded around the story of how you destroyed their home.*

STRATEGY #2—SPENDING YOUR WAY TO THE POOR HOUSE

Just because you spent money on something that was sold to you as a marketing program does not mean that it is worth the money you spent. The graveyard of home improvement contractors is littered with the corpses of those who spent a ton of money but did not get the jobs they had hoped for.

> *Just because you spent money on something that was sold to you as a marketing program does not mean that it is worth the money you spent.*

Terms You Need to Understand

Cost per Lead

Cost per lead is the actual dollar amount it ends up costing you to get a single lead from a marketing effort. For example, if you spend $2,000 for a particular marketing effort sold to you by a slick salesperson and it generated 40 leads, then the cost per lead is $50.

Marketing Cost per Job

If you close one out of 10 of those leads, your marketing cost per job is $500. This would be terrific if your average job netted $6,000, but would be upside down if your average job size is $300.

"Danger, Will Robinson!"

Merged Mailers

Merged mailers are marketing ploys that gang a bunch of different advertisements into a single mass-mailed envelope.

The merged mailer pitch sounds great. "Get your flyers into the hands of many qualified homeowners for a very low price." Merged mailers, in fact, are the most ignored form of junk mail. And we all know how often typical junk mail goes directly into the trash. Merged mail packets are worse. Think about your own home. How much junk mail do you receive every day?

Your advertisement is not just a single stand-alone piece of junk, it's hidden in an envelope (that has to be opened up) mixed in with 25 more pieces of junk.

Remember cost per lead. Merged mailers have a terrible cost per lead and are not worth your advertising dollars, even if the salesperson is incredibly slick. These salespeople are also notorious for telling you that you need to buy their program for multiple months (getting your name out there the infamous 29 times) in order to see results.

With all of that said, if you have a very large average job size and you have a big part of each job's profits set aside for marketing expenses, then perhaps this form of marketing will work for you.

Yellow Pages

Yellow page advertising is another perennial favorite of home improvement contractors. For years, having a yellow pages ad seemed to be a "must have" for home improvement contractors. It was perceived as a sign to others that you, too, were a significant player in your market by showing that you could afford a sacred yellow pages ad.

Now let's take a look at the logic. In today's world, who uses traditional printed yellow pages?

Prior to the advent of the Internet, most people used the yellow pages as a vehicle to find a contractor. Today, though, most people go to Google® or another search engine and type in "plumber in _____" or "handyman in _____." This gives them a list of all the internet-savvy contractors in the area as well as actual reviews about these contractors.

In today's world, who uses traditional printed yellow pages?

Customers can then jump on the contractors' Websites and make well-informed decisions as to who to call. This search method is quick, easy, and thorough.

Let me illustrate. Today, I drove by the mailboxes in my neighborhood where people of neighboring households go to get their mail. You know the type. A group of about 16 mailboxes clustered together on a single pedestal. Stacked underneath the cluster are several yellow pages books. They have been sitting there for about a month. Obviously, many of my neighbors don't see the value in bending over to pick up their copy of the local yellow pages book.

For a moment, think about a yellow page ad from a cost perspective. In today's day and age who would ever commit to an up-front astronomical monthly fee in order to hope to get a lead or two at a price point that no one can commit to? Don't do it.

This tip alone will save you many times the cost of this book!

I speak from painful experience. Years ago, I spent tens of thousands of dollars on paint contractor yellow pages ads. I received a call here and there, but once I started evaluating the cost per lead, I dropped out of the large-scale yellow page advertising program.

Do yellow page ads ever work?

There are some home improvement contractors who may still see some value in yellow pages advertising. These types of contractors would be the "crisis" type contractors. A crisis-type contractor is a contractor who is called immediately when an urgent problem occurs.

Do yellow page ads ever work?

The two types of contractors who best fit this category are roofers and plumbers. When people are in a crisis situation, they tend to revert back to simple methods they know and understand that have worked in the past. If their roof is leaking or their toilet is overflowing, they may rush to what is closest and makes them feel more comfortable. For many, that very well could be the yellow pages book.

They will then quickly peruse the ads and choose the one that looks like it offers quick and reliable 24-hour service. With list in hand, they call and call until they have someone . . . anyone . . . who can be at their house as quickly as the pizza delivery guy.

There are rarely painting, landscaping, etc. emergencies that will throw people into such a panic that they revert back to the yellow pages.

You may also benefit from the yellow pages if you live in an area with a very high elderly population that has a small local version of a yellow pages book. Be careful, though. Even the elderly are becoming computer savvy.

There are rarely painting, land-scaping, etc. emergencies that will throw people into such a panic that they revert to the yellow pages.

Summary of Strategy #2

- Be very, very, very cautious of marketing salespeople who want to take your money, but cannot guarantee you a cost per lead or, ultimately, a cost per job that makes sense to you.
- TEST everything you do in very small dollar amounts. Make sure it works before you commit to spending your mortgage payment.
- Understand your marketing cost per lead and cost per job.
- Most importantly, understand your average job size and what makes sense to spend to find a new customer.

TEST everything you do in very small dollar amounts.

STRATEGY #3—DOING STUFF THAT WORKS
Everyone Needs Your Services

The great thing about being a specialty contractor is that in almost every neighborhood in North America, almost every home on any given day has a list of things that need to get done that no one is getting to.

This "to-do" list was created by someone for someone else to do. Chances are neither of these people is eager to get his hands dirty to do what is on the list. The creator of the list is frustrated because the list is not getting done. The new owner of the list has less of an interest in getting to everything on the list than he or she does in cuddling on the couch and watching *Dancing with the Stars*.

The list is a treasure trove of potential jobs for any and every contractor.

If, for example, you are a painting contractor, the list has at least one room (soon to be a whole house, if approached properly) that needs to be painted.

If you are plumber, there's going to be at least one leaking fixture or plumbing problem that needs fixing on that list.

If you are an electrician, there is on the list at least one dimmer switch and light fixture sitting in corner of the garage that needs to be installed.

If you are a landscaper, every list has broken sprinkler heads on it. (And no one is ever happy with their maintenance landscapers . . . just ask them!)

Perhaps you are a handyman? The entire list is yours.

This isn't about fixing one leaky faucet or painting a small room. You have five guys who need to be put to work. The key here is to get in the door. Starting a conversation gains you access to that treasure trove list. On that list will undoubtedly be a larger project that needs to be done. (Pay particular attention to the section in Chapter 5 on "up-selling.")

The point here is that almost every home in America needs your services. The task is to find these people who need you through a marketing program that works.

> *The task is to find these people who need you through a marketing program that works.*

WHAT TO DO NOW—YOUR NEW MARKETING PROGRAM

Here is the section you have been waiting for. The section that gives you a leg up on your competition and saves you money!

Inventory List

Go and get everything that you use to promote your company. This includes everything from business cards and envelopes to flyers and advertisements.

List of what you have:

1. _____
2. _____
3. _____
4. _____
5. _____
6. _____
7. _____

Everything You Use

Everything you use to promote your business must have the same look.

Everything must have:

1. Logo and tag line
2. Phone number
3. Website
4. Email address
5. Contractor license information (depending on your state)
6. Colors that are all the same on each piece

Marketing Materials—What You Really Need

What marketing materials do you really need to get some business on a fairly regular basis? They don't have to be incredibly expensive.

Business Cards

The simple business card. The most important thing to remember about business cards is that their purpose is not as a tool for writing quotes on the reverse side! They are to be freely handed out to anyone and everyone who will take one. Better yet, they are to be given 10 at a time so people can hand them to their friends. If you are not going through 500 business cards a month, then you are not actively marketing your business!

> *The most important thing to remember about business cards is that their purpose is not as a tool for writing quotes on the reverse side!*

Your Basic Flyer

A basic flyer is the workhorse of your operation. It should be half of an $8^1/_2 \times 11$ sheet of paper (so you can get two out of each sheet). It should have all of your branding information that you came up with in the previous chapter, as well as any logos (with permission, of course) from any trade groups (including the BBB) to which you belong.

The "I'm Working Next Door" Flyer

This flyer is the same as the basic flyer except that it has a spot on it to put the address of the home in the neighborhood at which you are currently working. This notifies everybody in the neighborhood that you're the one who is doing the work down the street. These neighbors may want to nab you while you're there to do something similar at their houses.

Lawn Sign

A simple sign tells every passerby who it is that is working at the property. Remember, in most cases, people will be driving and probably paying more attention to the road than to the houses he is passing by. You want your lawn sign to be so easy to read that it communicates to the passersby—at a glance—what service you provide and how to contact you.

The basics of a good lawn sign:

1. Your service in BIG BLOCK letters:

ROOFING

PAINTING

LANDSCAPING

CABINET REFINISHING

Keep it short, using one- or two-word descriptions. This should take up 40% of the sign space.

2. Your phone number in BIG BLOCK numbers. This should take up 25% of the sign space.

3. Your Website in BIG BLOCK LETTERS. This should take up 15% of your sign space.

4. Everything else—your logo, license, anything else you want on the sign can fill up the remaining 25%.

5. Remember not to make it too busy or someone will drive right by and not know if you are doing flooring or pool cleaning.

Courtesy of College Works Painting.

Refrigerator Magnets and Other "Swag"

Swag has its place and can be helpful. If you happen to be in a service that a customer needs regularly, then something like a refrigerator magnet, a calendar, or anything along those lines that is likely to be saved by the customer has value.

If you are a roofer and you just put on a 20-year roof, why would you give anyone a one-year calendar with your name on it? Pens that are cheap will only associate your company with something that does not work and needs to be thrown out. Squeeze balls get lost or given to the kids. (Think they'll still have it when they are homeowners?) Pads of paper get used up eventually.

The best and least expensive item is a simple business card or two. Politely suggest they pin one up on the cork board that most homeowners seem to use to collect and keep their important information.

The cork board at my house has the business cards of my plumber, handyman and pool service guy. When I need those contractors, I know exactly where to find them. All the other goodies well-meaning contractors give me tend to go to the kids, the dogs, or in the trash.

Where do you get all of this done at a reasonable price?

Getting all of these various marketing materials produced is a vital part of your marketing plan and your budget. Do some research or ask other business owners about reputable companies that will produce all of this for you at a reasonable price.

> *If you are a roofer and you just put on a 20-year roof, why would you give anyone a one-year calendar with your name on it?*

Ads In Local Papers

I am generally not a fan of print advertising, but there are times when it can have an impact and not be very expensive.

Small local newspapers tend to be read cover to cover by people who want to know what is going on in their community. These papers generally have a very small distribution, but they do have a classifieds section or a services section that many homeowners turn to in order to find a local home improvement contractor.

Just like with anything else, test first and watch your cost per lead and cost per job.

Free Publicity

Nothing is more effective than free local publicity. An article written about you and your company in your local newspaper can bring you a flood of leads that will turn into jobs. And, best of all, it's free! Picture this . . . .

> *Nothing is more effective than free local publicity.*

Local Contractor Helps Needy During Tough Times

Harold Plumber, owner of Harold's Plumbing here in Colorado Springs, doesn't just talk about helping out his neighbors during these challenging times; he actually does something. Mr. Plumber and his employees took a Saturday afternoon and helped out a family here in Colorado Springs. John Smith and his wife Abigail are both retired and living solely on their monthly Social Security checks. What savings they did have were all but washed away in the stock market. Their home of 40 years had some significant plumbing issues and the Smiths needed help. Mr. Plumber came to their rescue after hearing about their plight on a local TV news report. After four hours of hard work by Mr. Plumber's crew, the Smith's dire situation was resolved. When asked why he did it, Mr. Plumber said frankly, "In today's challenging times, neighbors have to pull together and help each other out. The Smiths are great folks who just needed some help."

There are those who think an article like this may result in people calling for more free work. Yes, some people may call you for free work, but no more than those who already call you expecting to get a job for half your rate! You will also get plenty of calls from people who read the story, are looking for someone in your trade, and would love to give the business to someone who is out there providing charity service to people who truly need it.

So How Do I Do This?

First find a needy family in your area and block off a day or so to help them. Make the arrangements and go to work! Make sure the family writes you a reference letter. Enjoy the day and take plenty of pictures. You and your employees are really doing something worthwhile.

Before you go over to do the work, contact your local newspaper. Talk to someone in the features department. Small community newspapers are always looking for an interesting "feel good" story of someone doing something to help somebody in need. With all the hardship going on in our communities, it is great to read a story about a local merchant volunteering to help put some "hope" back in somebody's life. When you call, make sure to let the reporter know that the purpose of your service project is to show the community that people are still helping each other out.

Small community newspapers are always looking for an interesting "feel good" story of someone doing something to help somebody in need.

The Results

If you do this properly, you'll see two results. First you will likely receive calls for some new business. Second, you will feel really good about helping out people who need it.

Reference Book

Some of the best marketing comes from showing potential customers samples of your work and what you're all about.

Creating a simple binder can help undecided customers to get a sense of who you are as a person and what you do, as well as to get a visual for how much experience you have and the quality of your work.

A simple three-ring binder that is kept up-to-date with the following:

Some of the best marketing comes from showing potential customers samples of your work and what you're all about.

1. One page about your company, when it was founded, and what you stand for (your branding message).

2. A quick note about you and your family, complete with your picture. Remember, customers want to feel safe about the person they are hiring.

3. Copies of your workers' compensation and insurance policies.

4. A few pages with pictures of smiling employees working. Maybe add a bit about their personal family life and comments from customers about them individually. Customers need to feel comfortable around not just you, but your employees as well.

5. Testimonials about previous jobs. Each testimonial should include:
    *   A note about the scope of the job.
    *   Before and after pictures.
    *   A reference letter from the customer.

6. Product information that is relevant.

If you don't already have this put together, this should be the first thing you work on when you finish reading this book.

This tool alone will make you money right out of the gate.

MARKETING—REFERRALS

Every job you finish should result in you getting another job!

If you have done a good job and made sure your customer is happy, then each and every job you finish should lead to obtaining another customer. This does not happen by accident, though. If you are to be successful in having each customer refer you to their friends and neighbors, you need to work the system!

Every job you finish should result in you getting another job!

Customer Service Survey with Referral Piece

Every time you finish a job you should hand the customer a customer service survey. This survey is to be filled out on the spot, not mailed back (because very few people follow through on getting the survey in the mail). Design the survey to remind the customer of all the things you did well on the job. Additionally, the survey calls for the customer to write down the names and phone numbers of four friends who may be interested in receiving an estimate.

Every time you finish a job you should hand the customer a customer service survey.

The following are some examples of questions for your customer service survey:

1. Did our estimator show up on time?
2. Was our estimator polite and courteous?
3. Did our estimator take off his shoes or wear booties in your home?
4. Was our estimate clear, concise, and professional?
5. Did our workers show up when they were supposed to show up?
6. Were our workers polite and courteous?
7. Did our workers treat your home with the respect it deserves?
8. Were all of your concerns addressed?
9. Were you 100% happy with our job?
10. Would you refer us to your friends and family?

If the answer is "no" to any of these questions, you have the opportunity to resolve their concerns, either by actually fixing the problem or by apologizing for anything that was not up to their expectations.

The purpose of this survey is to simply reinforce, in a very positive way, how great your company is, what a superb job you just did, and—most importantly—to set them up to give your referrals. Your competitors are not doing this! This survey alone makes you stand out from the rest.

Referral—Call to Action

Your customers are very busy people. In most cases, even if they loved you and the job you did, they will not think on their own about referring you to their friends or take the time to do so. You need to entice them to help you out.

One great way is to offer your customers a reward for introducing you to your next customer. This can be as simple as a $20 Starbucks® gift card. In today's economy, you will be surprised how hard someone will work for you so that they can get $20 worth of coffee. Many of your customers have already cut out things like premium coffee from their daily lives. The reward, even a minor one, will be even that more enticing.

Take note, though, that in some states, such as California, offering a referral fee of any kind to your customers is illegal. Check with your state's contractors' board before offering any referral incentives. Where a referral fee is illegal, still go after the referrals by continuing to ask for them by calling to follow up with your customers a week or so after the job. In many cases, all it takes is reminding your customers that you are looking for referrals and they'll respond. Those who are really happy with you and your work will give you names simply out of a desire to help you out because you did such a good job!

Check with your state's contractors' board before offering any referral incentives.

Referrals 2.0—Facebook

Facebook is an overlooked marketing tool. Have your happiest customers put a quick recommendation of you on their "wall" on Facebook®. Facebook people are addicted to the site.

Facebookers do three main things:

1. Collect new friends.
2. Update their "wall" to tell their friends what they are doing.
3. Get constant updates from their friends about new things in their lives.

Facebook friends trust each other. Don't think that teenagers are the only ones who use Facebook—there are countless adults who use Facebook as a primary tool to communicate with friends near and far. Asking each of your satisfied customers to just post a quick, "Had my _____ done and so-and-so did a fantastic job!" will get you referral after referral.

BACK TO THE BASICS—PRE-INTERNET

Actual old-fashioned handwriting has its purpose!

There was a time when people routinely took the time to handwrite a note or a letter to offer thanks. Back 25 years ago, people handwrote them because not everybody owned typewriter and very few owned a personal computer. Email was definitely not mainstream, let alone a common practice.

Over the last 25 years, the art of the handwritten note has slowly disappeared. When was the last time anyone sent you a handwritten note or letter? Check out last year's group of holiday cards and note how many included handwriting.

If you really want to have a huge impact on your customers, send them handwritten notes thanking them for their business or the time they took with you on the estimate. (Many will stare at the card, reading it over and over again wondering where you downloaded the interesting font!)

The handwritten note will have impact. You will be remembered as one of the few that took the time to go the extra mile. And, with the note, don't forget to include a few business cards in the envelope for them to pass out to their friends.

The handwritten note will have impact. You will be remembered as one of the few that took the time to go the extra mile.

Use handwritten cards to:

1. Thank someone for his time after an estimate.
2. Apologize for a problem on the job.
3. Thank someone for his business at the end of a job.

For those of you who want to go on a limb, send them a $20 Starbucks gift card in advance of their first referral, letting them know that you are confident that they will more than earn the gift card (of course, only where this is legal). You will be surprised how many people will feel the need to make sure you get a referral because you already paid them for it!

Send Out Cards—Almost as Good

For those of you who have tried and tried to get on the program of sending out handwritten cards and have continually failed to do so, shame on you (OK, and on me, too).

I can honestly say that a handwritten card to thank a customer or an employee is the best way to make a substantial impression on your customers and your employees.

There is an easier alternative, though.

Sendoutcards.com is an online service that simplifies the process. The site has you input all of your customers' and employees' information into it. You pick the appropriate card and type a note on it. Another online card service, Senditout.com will print the card, put it in an envelope, and mail it out for you—all with the click of a button. The site also includes a handwriting feature that will copy your handwriting (or printing) so that the card actually looks like it came from you. What will they think of next?

EMAIL NOT FOR A THANK YOU, BUT . . .

Email simply does not have the same personal touch as cards and notes. Most working people today will receive 20 to 200-plus legitimate emails a day. The majority of business-people view email as a necessary evil. They see it as more of a challenge to their daily life than as a savior. Your message of "Hey, thanks for your time" will be forgotten almost immediately.

There is, however, a useful purpose to email.

Responding to a Customer's Questions

Always respond immediately to a customer email. (Yes, as if you were sitting around the computer just waiting for his message).

Even if your answer is "I will call you in 30 minutes" or "I will have to get back to you later today," you need to always respond immediately. (See the email section in Chapter 10 if you are wondering how to do this.)

Always respond immediately to a customer email.

Sending Directions to a Reference Job

When you give a customer a couple of your references' addresses to drive by to see samples of your work, always take the time to email an Internet map of the directions starting from their home. Even if they have a GPS in their car, they will appreciate the thought.

Emailing Reference Information

It is always a great idea to email to potential customers pictures of similar jobs and scanned copies of the reference letters that go along with them.

Emailing Information

Taking the time to email your customers information about your suppliers, products, and your Website link will really make an impression on those who are data driven and want to know more about you and your operation.

Extra Credit

If you really want to make a lasting impression on a customer, listen and have a conversation about their hobbies, interests, and diversions. Take the time to search out a couple of Websites or articles about one of their hobbies and send them the links.

You will make an amazing impression on them!

ALL YOU NEVER WANTED TO KNOW ABOUT THE INTERNET!

Yes, you need a Website. It does not have to be fancy. It does not need to be expensive. It can be done by any teenager in your neighborhood and it can be anywhere from one to 10 or more pages long. It really depends on how ambitious you are. There are, however, a couple of things to remember when setting up your Website.

The Why? of the Website

A Website is not going to suddenly drive new business to your door. Truth be told, when you put up a Website no one may ever find you unless they are specifically looking for your company. Try typing in "plumbers" or "painters" or "landscape contractors" into a search engine and you'll see pages of listings. Your chances of finding your way to the front page of such a generic search are almost as good as winning the lottery. (This despite what SEO [Search Engine Optimization] people tell you—but more on that later.)

Your Website's main purpose is to act as an online brochure about your company. Your Website is a visible and effective place for you to post photos, etc. of your most recent jobs. Customers can anonymously go to your site and get a better feeling of who you are when they are considering you work on their home.

Your URL (Otherwise Known as Your Web Address)

Ideally this is yourcompanyname.com. Depending on the unique nature of your business name, this may be very difficult to obtain. The best way to check the availability of a URL is to use Godaddy.com, where a feature exists that allows you to search for various versions of your name until you find something simple that works (and hopefully uses the .com suffix instead of .net, .biz, or the other extensions).

It is always a great idea to email to potential customers pictures of similar jobs and scanned copies of the reference letters that go along with them.

Your Website's main purpose is to act as an online brochure about your company.

The chance that your first pick is available is probably next to zero (unless the name of your business is incredibly unique). If you have not already secured yourname.com years ago, it has probably been taken by someone else. Take a cue from the movie industry. Have you ever noticed that every movie has a Website, but it is always _____ themovie.com? The movie industry realized that it would be next to impossible to always get the URL that it wanted. It understands the value of the .com extension, so it generally uses _____themovie.com to ensure nobody else has it.

If you can't find yourcompanyname.com, try adding your trade to the end. For instance, Builtrighttheframers.com or Smiththelandscapers.com or Joetheplumber.com. (OK, so Joe really doesn't have a business, but I couldn't resist.)

Once you have your URL, you will need to purchase the .net, .biz, and .org extensions, too. It is better to be safe than sorry. After all, once you implement the strategies in this book, you are going to be on your way to developing a large business—you might as well be prepared for the future. Also these extensions will only cost you about $10 each per year. That's not much of an investment to make certain that a competitor doesn't grab a similar URL.

Don't forget to renew your URL yearly or to buy the three-year package. Most URL providers allow you to "auto renew" so that you don't run the risk of losing the URL. Letting your URL expire is disastrous and needs to be avoided. An expired URL can be purchased by somebody else and you'll lose it. Confused customers trying to find you will fail and tons of printed marketing material is wasted!

Take a cue from the movie industry.

Don't forget to renew your URL yearly or to buy the three-year package.

Less is More

A picture is worth a thousand words. This old saying holds true on the Internet as well. In fact, if your front page has no pictures and just a thousand words, no one will stop to read it or even look at it. They will go somewhere else. In the Website game, less is more:

- Less long paragraphs (that people won't read) and more bullet points that jump off the page.
- Less colors and more white background and white space.
- Less words in general and more pictures.
- More pictures of you, your family, and your workers are great! Remember, people come to your site to get comfortable with your company. They want to feel good about you and your people coming into their home. The more your site can make them feel comfortable with the hiring of your company, the better.

References

A reference section is the one section that is a must for your site. Continually update the section with pictures and letters from your customers. There is nothing better than a site that shows work completed in the last 30 days. Customers tend to be suspicious of a company whose site only shows references from five years ago.

A reference section is the one section that is a must for your site.

Important Website Tips

Your Website should be simple and uncluttered. Remember, less is more, but some things must be included.

- **Contact Information:** This should jump off the page at your customers. Don't make them hunt for it. When they click on the contact info button, it should take them to a clean, uncluttered page that has all of your contact information, as well as a form to fill out if they would rather email you than call you.

   Your contact information should be big and bold at the bottom of every page as well.

- **"Free Estimate" Button:** Each page should have a "free estimate" button. This is yet another way to get to the contact information page.

- **Your Branding Message:** Your branding message (the one you came up with in the last chapter) needs to be front and center on your site. This is who you are and what your customers have told you is most important about you. This message, communicated in bullet points, is a key part of getting someone to choose you!

- **You Only Have 60 Seconds to Get Their Attention:** You all know how this works. You go to a Website and in the first 60 seconds you decide if you are going to spend more time there or just go somewhere else. Your Website needs to grab your potential customers' attention and speak directly to what they are looking for. If it doesn't, they'll quickly move on to a competitor's Website.

- **First Impressions:** In most situations, your Website is your customer's first impression of you and your company. If your Website is crowded, dated, or just plain boring, then your customer will naturally assume that is the kind of work you do.

> *If your Website is crowded, dated, or just plain boring, then your customer will naturally assume that is the kind of work you do.*

Blogs

Think about creating your own blog. Go to blogger.com and follow the easy instructions to create a blog. Blogs are a great way to keep your potential customers updated on jobs in progress and references in general. You can post reference pictures and letters from past customers and have general updates on things going on in your company.

> *Think about creating your own blog.*

Keep in mind that your company blog is not the place to share your personal, political, and religious views or your thoughts about the latest celebrity "it" girl. This is a space dedicated to just how cool your company is.

Just because you may have never read a blog and have absolutely no intention of ever reading one does not mean that your customers don't read blogs. Many of your customers obtain a great deal of their daily information not from newspapers or television, but from their Facebook account and from miscellaneous blogs on the Internet. And this trend is on the rise.

Your blog should have a direct link from your site and also be listed on all your marketing materials.

Once you start your blog, make sure you are committed to updating it at least weekly. People expect to see recent information on a blog. That is a blog's purpose.

Website Optimizing and Other Expensive Hobbies

In recent years, Website optimization has become a new up-and-coming business. The minute you launch your site, you will be pitched from multiple companies that will claim to be able to make you show up on the front page of a Google search. These salespeople will tell you that they have a service that, for a fee of thousands of your dollars (and many months of patience), can get your site to show up on the search engines. They are right about one thing. It is very expensive and it does take months. But there still is no guarantee that your site will show up anywhere close to the front page.

SEO—Search Engine Optimization

SEO is all about getting Google and Yahoo® to recognize your site as being relevant to the search that someone performs. This is done primarily by using keywords that Google recognizes on your site and then deems your site to have more relevance than another site.

There are challenges with SEO (not the least of which is whether you really need it). The main challenge is that Google does not recognize images as being relevant, so the more text you have on your home page with more relevant "keywords" and terms, the better your Website will rank. The challenge here is that you may rank high on the search results but, if you agree with the "60-second rule," people may actually click on your site and then disappear just as quickly as they got there because of a lack of interest in a home page flooded with text and no pictures.

No One Really Knows, Besides Google

Beware of any SEO company that claims it can guarantee you top rankings on Google. The algorithms that Google uses to rank its search results rival the secrecy of the Coca-Cola formula and Kentucky Fried Chicken® secret recipe. No one really knows what guarantees ranking success. And as soon as an SEO company starts getting some success, Google changes the parameters. It's an ever-changing game.

Beware of any SEO company that claims it can guarantee you top rankings on Google.

The $10,000 Question

The real question you need to ask yourself is this: do you really need to show up on the first page of a Google search? The majority of home improvement contractors operate on a local level. If your business is in San Diego, is it valuable to you that a person who searches for "plumber" in Boston finds your site among the search results? Let the big national players spend the money and worry about what happens when somebody types "landscaper" or "house painter" into a search box.

Searches That Do Matter

People have become much more skilled at how to find the information they are searching for on Google, Yahoo, MSN® and all the other search sites. Most people now will type in "handyman in _____" or "remodeling contractor in _____." By putting in their local city, the search engine will only come back with local results.

Own Your Local Searches

Owning the local searches is where you want to focus all of your Internet energy. Whenever someone types in your trade and the cities in which you operate, you want to make sure that your company comes up front and center. This is how you drive leads to your business. Keep in mind, though, that this is an expensive project if you plan on using SEO.

Much easier, less expensive, and guaranteed solutions exist.

Google—The One Stop Shop For All Your Web Marketing Needs

There are many different search engines out on the Internet. Some people have their personal favorites, but Google owns 67% of the Internet search market, followed by Yahoo at 20%, and the rest make up the remaining 13%. If all you do is make sure that you are set up properly on Google, then you will reach 67% of the people looking for your trade in your area. At a certain point, trying to figure out how all the other search engines work becomes a project with diminishing returns. Focus on Google local search and capture two-thirds of the market looking for you. Let the other third go—there are just not enough hours in the day to try and make all that work worth your while.

Online Maps

Utilizing online maps is an important part of your strategy. Make sure you are working with a company that understands how to use this technology to improve your local search results.

Utilizing online maps is an important part of your strategy.

Google AdWords

A great way to show up on the first page of a local search is to "buy" the spot on the Google home page. For what can range from pennies to dollars per "click," your business can show up every time someone searches for your trade in your area.

What is "pay-per-click? Pay-per-click means that you only pay when someone clicks on your ad. This is the ONLY way in which you should do Internet advertising. This is a strictly pay-for-performance model. If no one clicks on your ad, you don't pay.

Not Being Number One is Where it is At

Don't get carried away with your AdWords. You don't need to be paying top dollar for the first spot on Google. That game is for people who did not read this book. Let them waste their money. There is usually a premium charge for being the one in the first spot of the search results. It is amazing how much the price drops when you are number three or four on the pay-per-click ads. People who are looking for a home improvement contractor will look at the first three to five ads. They click on the links and go to the respective sites. At this point, they will decide from whom they want to get an estimate. (Go back to earlier instructions on how to build an amazing site!) You can end up paying $1/20_{th}$ the price that your competitor paid who bought the number one spot—and the potential customer ended up viewing both of your sites.

What it Boils Down To

The easiest way to make sure you show up when someone searches for "your trade" and "your city" is to buy the key words that will guarantee your business will show up each time that search is performed.

With a little extra reading, you can figure this out for yourself. If this is how you like to spend your spare time, check out www.perrymarshall.com. You can learn everything you ever wanted to know about AdWords at his site. While this is not a recommended task for you, the CEO of the company, it may be a good project for someone in your office.

Using a Web Marketing Company

For those of you who have no interest in learning an entirely new business, hire a Web marketing company to do it for you.

Finding a search company that understands your need to drive local traffic is the most important thing you can do with your online marketing money. Interview them thoroughly and make sure that you call at least three or four references. These references are vital in determining if their customers have been successful winning the local search battle.

One of the most important things to find out is how much of your money will go to the marketing company versus for your actual ad words.

This business sector is full of new companies that just started their businesses yesterday, so do your research!

My Website has my recommended companies for you as well.

> *For those of you who have no interest in learning an entirely new business, hire a Web marketing company to do it for you.*

Start Small

When setting up your AdWords account, start with a $100 limit or so and test, test, test. Monitor how many people are clicking on your site, and see and if this is indeed turning into contacts and estimates.

YES, THERE IS MORE YOU CAN DO ON THE WEB

Craiglist.org

Craigslist™ can be a tremendous source of leads. It is a relatively new phenomenon and is the "go to" place for almost any type of service (unless you have been living in a cave, you've heard of craigslist!)

- **Posting is easy:** Go to craigslist.org and click on "post to classifieds." Then click on "services offered" and finally on "skilled trades." At that point, just follow the directions. Make sure you are careful to use your "tag line" and put your company in the best light. Put a phone number that someone will answer and will not go to voicemail. Don't forget your email address as well as your company Website.

- **Post in every city you have coverage:** Craigslist covers very specific geographic areas so be sure to post your ad in each city in which you do work.

- **Repost every three days:** Craigslist is just that, a list. The key with any list is to keep yourself on top of the list. The only real way to do this is to keep posting your ad every three days so that you stay on top. This process only takes four or five minutes. There should be no excuse for not doing it.

- **Don't give up:** Over time you will get leads from craigslist. The key to success is to keep updating your entry in craigslist as part of your weekly routine. Write it in your calendar and follow through!

Craigslist covers very specific geographic areas so be sure to post your ad in each city in which you do work.

Backpage.com

Backpage.com™ is a competitor to craigslist. Take the time to post your listing with them as well. The process is the same.

CHAPTER 3

WHEN THINGS GO VERY BAD

In the old days, contractors could survive even though they did poor quality work and essentially left a wake of destruction and angry customers behind them.

Ten years ago it was literally possible to do just enough to get by. "The customer is always right" was a phrase that people would say, but only halfheartedly believed.

Many contractors were more likely to live by "Who are they really going to tell?" or "They never should have talked to me that way. Screw them. Let them clean up the mess." For years, this was the way that many of our peers, and even some of us, thought from time to time.

Truth be told, 10 years ago a home improvement business could actually grow over time and not take care of their customers. There wasn't much concern because the dissatisfied customers didn't have a way to get their irate voices heard.

TODAY IS DIFFERENT

Now it is simple for a customer who is not happy with your services to get the word out. Not only will they tell a couple of friends and neighbors about how dissatisfied they are with you, but also they'll tell the whole world.

And damage isn't done with just the completely unhappy customers, but with customer comments that the job was only "satisfactory," not "outstanding." In today's world of customer-driven review Websites, "satisfactory" is about the same as "Oh, and they killed my cat as well."

SATISFACTORY WILL KILL YOUR BUSINESS!

What is wrong with doing a satisfactory job? You'd think there would be nothing wrong with a satisfactory job, but it simply isn't enough in today's electronic age. "Satisfactory" today means you are not planning to grow your business or maybe not even planning to stay in business.

Be aware: customers will research you on the Web!

This is not 10 years ago. Wake up! Check out your local competitors. Consumers now have a wealth of knowledge available to them on the Web. Consumers will spend hours and hours researching and reading the reviews on different sites before they take the plunge to buy a $200 digital camera. How much research do you think they will do

Be aware: customers will research you on the Web!

before they give thousands of dollars to a home improvement contractor who is going to be inside of their home with their family?

Chances are they will spend four or more hours doing their homework on the Internet before making a final decision on who to hire.

Their research has been made easier by a brand new group of entrepreneurs whose Websites are driven not just to make money, but by social responsibility. One of their main purposes is to make sure that home improvement contractors are held to a higher standard than they have been held to in the past. These sites do what contractors' boards across the country have been trying to do for years.

Their aim is to scare you straight. And it's working!

One of their main purposes is to make sure that home improvement contractors are held to a higher standard than they have been held to in the past.

THE SITES AND WHAT THEY SAY ABOUT THEMSELVES

Pissedoffconsumer.com™

"The Pissed Off Consumer! Pissedoffconsumer.com is the best consumer product review and complaint site online. Have you been scammed, cheated, received terrible service, or something similar? Post your complaint and share your opinion feedback with site members about products or services. Get the word out today and have your voice heard!"™

Angieslist.com™

"Don't know who the best—and worst—contractors are in your area? We do.

Your home is your most valuable asset; don't let just anyone work on it. Angie's List is a word-of-mouth network for consumers with hundreds of firsthand reports from members in your area on who they recommend you use or who to avoid."™

Ripoffreport.com™

"Are you a victim of a consumer rip-off? Do you want justice?

Ripoff Report® is a worldwide consumer reporting Website and publication, by consumers and for consumers, to file and document complaints about companies or individuals. While we encourage and even require authors to only file truthful reports, Ripoff Report does not guarantee that all reports are authentic or accurate. Be an educated consumer. Read what you can and make your decision based upon an examination of all available information.

Unlike the Better Business Bureau, Ripoff Report does not hide reports of "satisfied" complaints. ALL complaints remain public and unedited in order to create a working history on the company or individual in question.

Ripoff Reports cover every category imaginable! You can browse the latest reports, search the reports, or submit your report now for FREE, by clicking on File Report. View over 1,000 different topics and categories.

By filing a Ripoff Report it's almost like creating your own Website . . . and it's FREE.

Your Ripoff Report will be discovered by millions of consumers! Search engines will automatically discover most reports, meaning that within just a few days or weeks, your

Your Ripoff Report will be discovered by millions of consumers!

report will probably be found on search engines when consumers search, using key words relating to your Ripoff Report.

Ripoff Report is here for you, the consumer.

Search the Ripoff Report before you do business with retail stores. Check to see if they have a history of bad return policies, checking and credit theft, rebate fraud, or other unscrupulous business policies, such as phony auto repairs, auto dealer bait-and-switch tactics, restaurants with bad service or food, corrupt government employees and politicians, police corruption, home builders, contractors, unethical doctors and lawyers, online stores that sell nonexistent products, deadbeat dads and moms, landlords and tenants issues, fraudulent employment and business opportunities, and individual con artists who scam consumers."™

Yelp.com™

"Local reviews of your service."™

Homestars.com™

"Read Reviews. Write Reviews. No membership fees.

Who to hire, where to buy, and who to avoid.

HomeStars enables homeowners like you to connect with neighbors online."™

Checkbook.org™

"Who's rated the best and the worst? Consumers' Checkbook is the independent, non-profit consumer authority. Consumers' Checkbook rates the local service firms you use in your everyday life. With this powerful information, you'll save money and time for things that matter."™

Kudzu.com™

"Another great site for customers to tell the world about their favorite or least favorite businesses. Consumers rate businesses from one to five stars and then review the businesses for others to use in their evaluation of potential businesses they may be thinking about using."™

Google®

"Google also has the ability for a customer to do a review on their local search sections, so when a customer looks you up online the good, the bad, and the ugly will show up! "™

SCARED YET?

After reading the missions of these various sites, if you're not just a little scared, then you're still not getting the point. The world has changed, and you can no longer hide from your faults or mistakes. These sites are real, and customers are using them more and more every day to help them make decisions as to who to hire and who not to hire.

After reading the missions of these various sites, if you're not just a little scared, then you're still not getting the point.

THE CUSTOMER IS NOT ALWAYS RIGHT!

You know I really do agree that customers are sometimes wrong.

I'm convinced that 98% of the time, customer problems are caused by something we the contractors did, but in 2% of the cases, the customer is the one at fault and is only trying to take advantage of the situation.

Up until about five years ago, we all knew exactly what to do with the 2%. Every home improvement contractor has her favorite stories of those 2%.

The challenge today is that even when consumers are dead wrong, you can't afford not to make them happy. That customer who did not pay you because of some bogus claim can still make your life hell. His access to all of these Websites will cost you much more business than just biting your tongue and giving in.

Unfair? Of course. But, trust me, even if you feel like you have won the battle by standing firm or not bending your ways, you will definitely lose the war.

Make them happy and move on.

The challenge today is that even when consumers are dead wrong, you can't afford not to make them happy.

What Would You Do?

All it takes is one or two upset customers to post something negative about your company on any of these Websites and your future business will start flying out the window.

How many customers will decide to call your competitor because you had a poor or average review and your competitor had a positive review? What do you do when you see negative reports about somebody you are considering doing business with? If you were researching companies on the Web and one business had negative or satisfactory reviews and the other one had great reviews, who would you choose?

The answer is simple.

How many customers will decide to call your competitor because you had a poor or average review and your competitor had a positive review?

The Review is Incorrect! I'm Suing!

The site is wrong. They had better take down the negative posting about my business! They can't do this! I will sue them!

You're not going to be successful in either getting the various sites to delete the negative review or in suing the Website owner. The best way to summarize why you will not win is to tell you about Ripoffreport.com's site. Ripoff Report™ has a section on their Website explaining why they never remove claims against company or individuals, even if these persons or companies believe or show the comments are incorrect and defamatory. Ripoff Report goes on to explain that deleting reports would threaten the credibility and accuracy of the Website. They also cite a federal law known as the Communications Decency Act or "CDA", 47 U.S.C § 230, and use it to support the argument that they are not responsible or liable for anything posted by an outside individual on their Website. So there you have it. If a negative review goes up on Ripoff Report's site, it stays up on the site forever. Even if it is a total lie.

The new rule in customer service is the same as the old rule.
THE CUSTOMER IS ALWAYS RIGHT.

Now, Do I Have Your Attention?

The new rule in customer service is the same as the old rule.

THE CUSTOMER IS ALWAYS RIGHT.

The only change in the rule is that you had better take it seriously. If you don't, you will eventually be out of business!

I HAVE A PROBLEM. NOW WHAT?

So somehow you managed to get yourself a bad review on one of these sites . . . now what are you supposed to do? The first thing to do is to take a deep breath and realize that this does not need to be the end of your business as you know it.

There are things you can do to actually help fix the problem.

First, grit your teeth and call the customer and resolve her concerns. This may be the last thing you may want to do at this point—especially if they are part of the 2%—but you just have to do it.

Call or visit them and offer to fix the problem. Give them all their money back if you have to. Do *anything* you can to get them to post another comment saying that you did indeed take care of the problem.

No one really knows how much business a bad review will really cost you! You have no idea how many people were impacted by it. Ignoring the bad review will only make bad things even worse.

Be Proactive

It is imperative to stay ahead of any bad press you may get.

The best way to do this is to make sure that you are offering the highest level of customer satisfaction possible. Above all, when something gets screwed up (and it will), apologize right away, fix the problem, and then apologize again.

Once you get home send the homeowner an apology note and a Starbucks gift card, and then reread the customer satisfaction part of this book. You simply cannot afford to have any bad press on the Web.

It is imperative to stay ahead of any bad press you may get.

Bad Review Half-Life

Question: What is the half-life of a bad review on the Web?

According to Wikipedia®, the definition of half-life is:
"The half-life of a quantity whose value decreases with time is the interval required for the quantity to decay to half of its initial value. The concept originated in describing how long it takes atoms to undergo radioactive decay but also applies in a wide variety of other situations."

Answer: The half-life of a bad review on the Web is just like throwing plastic into a landfill. It almost never goes away.

A bad review on the Web will be around as long as the Website is around, and just like a fine wine, it gets better and stronger with age! A bad review on the Web grows stronger and stronger the more often people look at it. It will move up the list when searched in Google®, etc. The more clicks it gets, the more relevant Google will deem it to be and the higher it will rank on Google's search results list. That 2% of people who used to just go away and disappear can now haunt you forever with one simple posting.

They may forget about you, but you will never forget about them.

A bad review on the Web grows stronger and stronger the more often people look at it.

That 2% of people who used to just go away and disappear, can now haunt you forever with one simple posting.

HE WHO LIVES BY THE SWORD WILL DIE BY THE SWORD

The moral to this section is that if you slash and burn your customers, then they will slash and burn you. They just carry a bigger sword and will do much more damage to you in the long run.

But this Customer is Dead Wrong!

We have all had the customer who is out-and-out trying to take advantage of the system. If you think we as home improvement contractors have it bad, spend a little time with employees from a Home Depot® or Lowe's® store, and listen to their stories about customers trying to push the boundaries of right and wrong.

The problem with drawing a line in the sand and taking a stand is that you really can't win anymore. This may sound like you should always just give up and not fight back, but the truth is that in today's Internet age it is much better to retreat so that you can live to fight another day. Fighting a fight based on principle, even when you are absolutely right, just isn't worth it.

> *Fighting a fight based on principle, even when you are absolutely right, just isn't worth it.*

The World is Generally Good

I tend to believe that most people are inherently good and are not out to take advantage of their fellow man.

I also believe that the small percentage of the world's population that is out to do harm and take advantage of people never end up doing well enough in life to own a home and therefore are not our customers.

I also believe that most of our customer challenges can be traced back to something that we did as contractors and therefore we're the actual source of the problem.

As for the little group who is truly evil and spends its time plotting against unsuspecting home improvement contractors, they'll get theirs. Karma is a wonderful thing.

> *I also believe that most of our customer challenges can be traced back to something that we did as contractors and therefore we're the actual source of the problem.*

WHAT YOU CAN AND SHOULD DO NOW

Constantly Google Yourself

Spend the time to go online and Google your company so that you see what comes up. Don't forget to search your company under all of the customer Websites and make sure you know what people are saying about you.

Encourage the Positive

Take advantage of your positive customer relationships and ask customers who really like you to go onto each of the sites and post a rave review of your company. Maybe even have an inexpensive business card printed with all of the sites' Web addresses on it.

Why can't you have 10 or more positive reviews on each and every site? All it takes is for you to take the time to ask people to help you out.

LEADS

There is no such thing as a bad lead.

The car business has a saying, "They don't walk onto the lot if they are not looking to buy." Essentially, this means that every customer that walks onto a car lot has an interest in buying a new car.

For what other reason would people go to a car lot? They could argue that they are "just looking," but deep down, they have an interest in that new car they're looking at.

If the right deal could be made, they would buy. That "right deal" might be $0 down and zero payments forever (or more likely something slightly higher than that), but in fact there is a deal out there that would make them buy.

Heck, if that car is the car of their dreams and the stars align just right, they may even be persuaded to pay full price and buy on the spot. Anything is possible. Those sales take place everyday.

Automobile salespeople are confident they "have you" the moment you walk onto the lot, because they know you were motivated enough to make the trip to their dealership to take that first step. The interest is there; and they know it!

Ever walk into a fast-food restaurant just to check out the prices?

People do things for a reason. We go into restaurants because we are hungry, we walk in to grocery stores to buy food, and we go to shoe stores to buy shoes.

EVERYONE WINDOW-SHOPS

Actually, no one really window-shops . . . they always end up purchasing *something*.

When was the last time you ever saw a person spend a day window-shopping and not buy anything?

If they are out wandering around the mall, they will buy *something*. It may be small, but they'll buy something. If not, they would have gone to the park or stayed at home.

HAVE YOU EVER BEEN TO MEXICO?

If you want to see the best example of believing that the world is your customer and that everyone will buy from you, all you need to do is go south of the border for a day or so.

Just walk around any open-air market and you will be accosted by vendors beckoning you to come into their shops.

If you happen to be wearing any type of identifiable clothing, they will address you straightaway. Your clothing logo becomes your name. "Hey, Oakley®! Come on over here. I have a great deal for you."

"Utah! Hey, Utah! Yes, you, Utah! Come in my shop. I have exactly what you are looking for. I'll make a great deal—real cheap—just for you, Utah."

You tell the aggressive vendor that you don't have time and that you're on the way to the beach. There is no obstacle large enough to keep him from making a sale. "No problem. I'll bring everything directly to you on the beach." He brings that and a whole lot more, including a bunch of stuff you never knew you wanted.

Only in Mexico will people sell you wool blankets on the beach in 95-degree heat! Why? Because the gentleman selling the blanket really believes that you came to his beach today looking to buy that blanket.

And guess what? In his world, he is right. If he works hard enough, talks to enough potential customers, smiles his best smile, and says the right thing to make you laugh, you'll end up looking at his blankets. If he gets you to respond to him, he knows that he has a 50/50 shot of making the sale. If he gets you to ask, "How much?" he's 80% there. All it takes to close the deal is the right price.

You came to Mexico and you walked onto *his* beach. You are his lead, and you are his customer. He just needs to cut you the right deal for the blanket that you will ultimately take home, put in a closet, and only get it out when the kids have a sleep over.

How many of you would choose to sell wool blankets on the beach in 95-degree heat? How many of you would even have a chance to succeed at it?

HOME IMPROVEMENT LEADS

In our world, a lead is someone who has expressed any level of interest in what we do.

A qualified lead is anyone who gave us a phone number or email address.

In our car purchasing case, in the eyes of the car salespeople, the customer not only walked onto the lot, he might as well have screamed out at them, "Hey, call me. I think I want to buy one of those."

The disastrous thing about our industry versus our aggressive "come-into-my-shop" vendor in Mexico is that when we have our potential customer "walk onto the lot," we don't act or speak until we make absolutely sure the lead is truly a *really good* lead.

Many of us first confirm everything about the lead prior to going to see the customer.

We tell ourselves that if we are going to take our valuable time and go to a customer's home, we want to make sure that they are very serious about buying today . . . not tomorrow, not in a week, but *today*. We make sure customers first convince us of their absolute sincerity to have the work done. If not, then we don't agree to show up for an estimate. And, if that isn't enough screening, we insist the job must be large enough to warrant our time and, of course, "will you be paying by check or charge?"

When most of us see someone "walk onto the lot," we ask for their social security number in order to run a credit check before we even see what kind of "car" they might want to buy!

The Leads Are Weak . . .

When you have some time, you need to rent one of the greatest sales movies ever made, *Glengarry Glen Ross.*

There is a fantastic line by Alec Baldwin in the movie when he says, "The *leads* are weak? *You* are weak!" This movie is a fantastic example of how most of us are always looking for the perfect "lay down." That lay down is something like the person who tells us, "Yes, of course, please charge my credit card for the full price of the job before you come over and see my house. I want to thank you in advance for taking the time to do the work for me. I will make sure my wife has lunch ready for you—what would you like to eat?"

HAVE YOU EVER?

Have you ever been absolutely sure that you were going to book a job on the spot? You know the jobs I am talking about.

The potential customers who tell you on the phone how they saw your last job and how great the work was. They are best friends with your last customer and, "Wow, did you ever make them happy! And, by the way, they hated their last four contractors. You really must be something special."

They want to know what your schedule is and how soon you can get started, even before you have met them. When you go to meet the customers, they practically give you a hug at the door. You present your estimate and they tell you how great it looks and that they just need to check over a couple of things, but starting in two weeks sounds about right.

You leave their house with a smile on your face and then cancel your next customer meeting because it was only for a couple hundred bucks worth of work. You decide to take the rest of the day off.

A week later, you can't get back in touch with them and you're not sure what is going on. You cleared your entire schedule to start next week, stopped pushing for more work because this job would "keep the guys busy for a month." But . . . guess what? You never talk to them again.

Have you ever? I bet you have.

Have you ever met potential customers who so rubbed you the wrong way that the only reason you even ended up doing their estimate was because you could not think of a good enough excuse to leave their home?

Have you ever met potential customers who you were convinced were made of stone or some other type of indestructible matter? Customers who believed that "stoic" is an emotion that is well-suited for all occasions? Customers where you notice even their own dogs don't like them?

Have you ever?

And then, have you ever been completely shocked when they grab the contract from your hand, sign it before you get a chance to go through it, and then ask when you are going to start?

I bet you have.

There is a fantastic line by Alec Baldwin in the movie when he says, "The leads are weak? You are weak!"

Have you ever met customers who swear that they only need a few things done and don't want to spend a bunch of money? Customers who start the conversation with, "I am thinking of doing this myself," or "Last night while I was changing my oil. . . ."

Have you ever?

And then, have you ever seen that customer end up deciding to do twice as much as you would have suggested be done and not flinch at the price?

I bet you have.

Have you ever had customers promise you that they were *absolutely* going with you, then when you drive by a week later there's someone else doing the work?

Have you ever had customers tell you that they were not going to start the work until the next year and then end up agreeing to start the job the following week? Have you ever had customers claim to want the cheapest job on the planet, yet sign for all the options? Have you ever had a simple estimate turn into a huge job you were not expecting?

Have you ever seen a customer with a brand-new BMW® or Mercedes® in the garage ask for the cheapest materials possible?

I bet you have. Everyone has experienced the same or similar situations to these examples. So why do most contractors still feel the need to keep "grilling" prospective clients?

> *Have you ever had customers claim to want the cheapest job on the planet, yet sign for all the options?*

The Lead Qualification Game

Over-qualifying a lead is a mistake that most of us are guilty of. I subscribe to the idea that if people give you their number, then they are looking to have some work done.

If they are looking to have some work done, then they all have the same potential. It is your job to unlock that potential!

> *Over-qualifying a lead is a mistake that most of us are guilty of.*

All the World Is a Stage . . .

Every customer is an actor. They act in the way they believe they need to act in order to get the best deal possible.

Some believe that the more friendly and positive they are, the better the deal they will get.

Others believe that if they are really tough and no nonsense, then *they* will get the best deal.

Some believe that if they act like they are willing to do it all themselves, then you will be forced to bring down your price in fear of losing the job.

What they all have in common is that at one time they all walked onto a car lot, went window-shopping, bought a blanket in 95-degree heat in Mexico, gave you their number, wanted some work done at their home, and they didn't really want to do it themselves.

SOME UNDENIABLE TRUTHS ABOUT LEADS
Good Leads Die Young

Prospective customers want you to call them back now, not later.

The best way to call the customer back is not to have to call them back at all. The ideal situation is that when a customer calls, you have someone in your organization who can take that call immediately.

> *Prospective customers want you to call them back now, not later.*

Think about it. When potential customers decide they want to talk to someone about getting some work done, they are in a buying mode. They have sat down, blocked off the time, are mentally prepared to speak about their project, and have dialed the phone.

They did not do all of that so that they can get a call back in four hours or four days. They're ready to talk now—on their schedule, not yours. They called so that they could talk to *you* about using *your company* for work on their home.

If you don't talk to them right away, what will they do? Will they wait for you to get back to them? Will they give up in frustration and decide not to get their project done?

No, what they will do is keep dialing the phone until they talk to someone. They will be at their best—most excited and most receptive—to the first person they talk to.

Their enthusiasm drops the more times they have to tell the story.

Old Leads Are Not Bad Leads

Another common misconception is that a lead that is a week or a month old is no longer good at all. Generally, most contractors will call on a lead once or twice. If nothing happens, they throw it in the trash and mutter something about people wasting their time.

A lead is only dead when the customer says they don't want the work done or you are too late and someone who was more persistent got the job.

A lead is only dead when the customer says they don't want the work done or you are too late and someone who was more persistent got the job.

If They Don't Call Back, They Are Not Interested

Another great misconception about customers is the idea that if they don't call you back, then they don't want the job done.

The truth of the matter is that you are not as important to them as they are to you. Customers are very busy. They are working more hours than ever before, balancing spouses and kids, activities, and whatever personal life they can fit in. Remember: they called you when they were ready to talk. Now, they are busy and will get back to you if and when they have the break in their schedule to do so.

This does not mean that they are not still interested in having the work done. It only means that they didn't have the time to talk to you at the moment you called so they didn't respond.

Last Man Standing

Persistence is the key to success when reaching customers.

Call regularly and at different times. Leave messages only once a day, but call more often. Mix up your messages. Make it lighthearted and fun. If you have their cell phone, send a text message. Don't stop calling until they tell you to stop calling.

In 90% of the cases, they simply have not called you back because they have been too busy.

When you actually do end up writing quotes for customers who required 10 phone calls, in many cases they end up booking with you. The reason why they booked with you is because all of your competitors gave up calling them back. They feel obliged to go with you after all they put you through to get a call back. Your estimate, oftentimes, is their only estimate.

Leave messages only once a day, but call more often. Mix up your messages. Make it lighthearted and fun.

EVERY LEAD IS DIFFERENT—BUT THE SAME

Leads will come from a variety of sources. They will come as responses to referrals, vehicle signs, your staff on a job site, a flyer in the neighborhood, your Website, a door you knocked on, or a lead you bought off the Internet. Each lead is unique and the first phone call needs to be tailored a little differently depending on the characteristics of the lead.

Characteristics of Leads

Referral

- Already has heard about you.
- Most likely going to have something done.
- Arguably the warmest of the lead family.
- Most likely not a competitive bidding situation.

When dealing with leads from referrals, you should:

- Answer the phone or call back right away.
- Talk up the customer they know.
- Get a letter of reference from their friend.
- Get out there ASAP to book the job.

When dealing with leads from referrals, you shouldn't:

- Take you sweet time getting back to them.
- Assume the job is yours at any price.
- Get lazy with your presentation because you assume it is yours.
- Overcharge because it is a referral.

From Your Crew

- Already has seen one of your jobs and likes what they saw.
- Most likely going to have something done.
- Likes your crew (they gave their phone number to a stranger).
- May or may not be a competitive bidding situation.

When dealing with leads from your crew, you should:

- Talk up your crew (remember, they like your crew).
- Talk up the job that they saw your crew doing.
- Have your crew call you while the customer is standing there so that you can talk to them right away.
- Get out there right away.
- Incentivize and thank your crew for every lead.

When dealing with leads from your crew, you shouldn't:

- Forget to sell them on you and the company.
- Take your time calling them to set up the estimate.
- Assume they won't forget who you are.

Flyer/Yard Sign/Truck Sign

- May or may not have seen your work.
- Most likely going to have the work done.
- Most likely knows very little about you.
- May or may not be a competitive bid situation.

When dealing with leads from flyers or signs, you should:

- Take the call right away or call back in five minutes.
- Find out where they heard about you so that you can see what marketing is the most effective.
- Ask a ton of questions to understand what they are looking to do.
- Educate them on you and your company so that they have confidence and are looking forward to meeting you in person.

When dealing with leads from flyers or signs, you shouldn't:

- Wait two days to call them back.
- Assume that they will or won't buy based on the phone call.
- Be arrogant or standoffish on the phone.
- Wait too long to set up the estimate—get out there right away.

Your Website

- Most likely has read about you on your site.
- They will do their research on you.
- Most likely a competitive bidding situation.

When dealing with leads from your Website, you should:

- When speaking to them, refer to your Website often and even get them to go online while you are on the phone to walk them through the highlights of the site.
- Recognize that they will do their research and will want lots of information.
- Take your time and don't rush to get them off the phone.
- Make sure they get all the information they need.

When dealing with leads from your Website, you shouldn't:

- Forget that your competitors' site is just a click away.
- Ignore email and texts.

- Assume the Website will take your place in the sales process.
- Assume that they are already sold.

Cold Call

- Most likely has never heard of you.
- May be just at the beginning phases of thinking about the project.
- You may have put the idea in their head.
- You may be the first and only bid, or one of many.

When dealing with cold call leads, you should:

- Set the date and time of the estimate immediately.
- Call back multiple times; people are busy.
- Educate the customer because they don't know who you are.
- Convince them to meet you.

When dealing with cold call leads, you shouldn't:

- Be put off if they are not as excited as you would like.
- Expect them to be excited to talk to you yet.
- Stop calling until they say no.

Store-Bought Leads

- Customer may or may not remember filling out an online form.
- They are now receiving calls, not in the mode of making a call.
- May or may not be ready to talk when you call back.
- This will be a competitive situation with three to eight contractors.

When dealing with store-bought leads, you should:

- Get the lead on a text on your phone.
- Call as soon as the lead comes through.
- Get to the customer's house for the estimate right away.
- Educate the customer on your company.

When dealing with store-bought leads, you shouldn't:

- Assume the customer knows anything about your company.
- Be put off by the fact that, at this point, you are just one of six people calling them.
- Give up on a lead until they say no.

Article in the Newspaper

- They already like you because of the article.
- You have a smidgen of "celebrity status."
- May not be a competitive bidding situation.

When dealing with leads from newspaper articles, you should:

- Talk about the article and what a great experience the event was.
- Bring a copy with you to the estimate.
- Take advantage of your celebrity status.

When dealing with leads from newspaper articles, you shouldn't:

- Assume that you have the job because of your article.
- Be overconfident just because you got some free press.
- Forget that fame is fleeting and this is your 15 minutes—use it wisely.

Yellow Pages or Newspaper Ads

- They are looking for a contractor.
- They know very little about you.
- They want you to answer the phone or they will call someone else.

When dealing with leads from ads, you should:

- Realize these buyers, more than likely, are attracted to doing business with a local company.
- Take the call right away or call back within five minutes.
- Ask what was it about your ad that made them call you.

When dealing with leads from ads, you shouldn't:

- Assume that your ad has told them all they need to know about your company.
- Think that they will remember calling you off of the ad if you call them back three hours later.
- Think they called only you.

The Do's and Don'ts

The do's and don'ts are about the same for almost every lead. You need to make sure you take customer calls immediately. Take the time to listen to what they want. Sell them on you, as an individual, as well as on your company and team. Continue to be responsive throughout the process.

If you follow these basic steps, you are already miles ahead of your competition.

> *Take the time to listen to what they want. Sell them on you, as an individual, as well as on your company and team.*

INTERNET LEAD PROVIDERS—A FEW IMPORTANT THOUGHTS

Over the past few years, Internet lead providers have popped up all over the Web. These sites, such as Service Magic™ and Reliable Remodeler™, spend hundreds of thousands of dollars advertising on the Web for people who want to have home improvements done.

There are now a plethora of sites where you can purchase leads. All you need to do is Google® "home improvement leads" to see just how many are really out there. If you Google "plumber," "painter," "handyman," "landscaper," etc., you will most likely see these lead providers show up on the first page of the search results. They will show up in the top 10 and probably on the side panel where the paid searches appear.

These sites are very powerful and gather thousands and thousands of leads for home improvement jobs.

Why They Are So Popular

Customers love these sites because they are able to go to one location, fill out one form, and get "prescreened contractors" to call them. They view it as a timesaving and quality control system.

With this service, they don't have to take the time to find quality contractors. The work has already been done for them.

The Customer Experience

The customer experience is quick and easy and mostly painless. That is, of course, unless they filled out a form on multiple sites. They'll soon be inundated with as many as 12 different contractors' calls.

Customers tend to be pleased with the experience, though, because they end up getting a very competitive bid by a qualified contractor.

The Internet Is Not Going Away

Everyone has the crazy old uncle (or maybe *you* are the crazy old uncle) who swore that the Internet was just a fad. The old codger was convinced that, soon enough, just like the pet rock, the Internet, too, would pass.

Well, the crazy old uncles are wrong and the Internet is only getting stronger. Websites like Service Magic and Reliable Remodeler are growing in popularity amongst consumers. If you don't take them seriously, you will be missing out on potential new business. You will end up being the guy you heard stories about growing up. You know the one. The one that believed that new fangled television thing was just another fad.

If You Can't Beat Them, Join Them

So now that you are a believer, it is important to truly understand how this game is played. Each site is a little different, but here is the basic premise.

Follow these basic steps to set yourself up to receive Internet leads:

1. Choose a site and click on the "Contractors Only" section.
2. Sign yourself up as a contractor.
3. Make sure you have a valid credit card to pay for your leads.
4. Provide whatever references and insurance documents that are needed.
5. Define the geographic territory from which you are willing to take leads.

6. Define the type of leads you are willing to take.

7. Set your preferences to receive the leads via email or cell text.

If You Are Not First, You Are Last

In the immortal words of Ricky Bobby from *Talladega Nights*, "If you are not first, you are last."

The most important thing to remember about these types of leads is that this absolutely is a race! Each lead is given out simultaneously to multiple contractors. They will all be attempting to contact the customer. If the customer has made requests on multiple sites, then he might have 12 or more contractors calling.

You want to be the first to contact the customer. You want to set the bar high enough that the others who will follow you won't be able to measure up.

Set an appointment to meet the customer in person as soon as possible.

Enjoy the Pain

I won't mislead you. These leads are not as fun to call on as those from people who are best friends with your last customer or who read an article about you in the local newspaper. They don't know you. You are going to have to earn their trust and business the hard way.

These are real leads, though, and will turn into real jobs if you work them properly.

Bad Internet Leads

From time to time you will come across a lead that is not legitimate. Maybe someone put in their friend's name and number just as a joke.

These leads won't cost you anything, though. You can return them to the vendor and get credit for these bogus leads. Keep in mind, though, that 90%-plus of the leads you will get off of these sites are real homeowners who really want to have the work done.

If you feel you are getting higher than a 10% "junk lead rate," then the problem is most likely you and not the lead.

The Secret

There are some home improvement contractors who do a substantial portion of their business through buying leads off these sites.

The vast majority of contractors will sign up for leads, take five or six, take their time calling them back, and end up not booking a single job. These contractors will then go back to sitting at home, hoping that the phone is going to ring.

The secret is in being first, not last!

Invest

When you sign up, realize that you will probably mess up the first few leads while you get accustomed to dealing with these types of customers.

Commit to taking 20 leads and aggressively work them. If you follow the steps you read in this book, you will have success with these sites. The other benefit you have is that your competitors have not read this book. They will end up taking five leads, will get mad, and quit.

All the more leads for you.

BRANDING/MARKETING/LEAD GENERATION SUMMARY

Branding

Figure out who you are and what is important to your customers. Make sure that you are aligned in this area and then create your commercial, tagline, and company brand. Your branding message will be the same on all your marketing materials.

Marketing

Getting your name out into the community does not need to be an expensive endeavor. It does, though, need to be a constant effort.

Lawn signs need to go up on all of your jobs. Flyers should be passed around the neighborhood you are working in. You need to always be passing out flyers and business cards. Seek press coverage of the great things you do. Make sure you obtain referrals from past customers. These activities all need to be part of your daily schedule.

Lead Generation

Cost per lead and marketing cost per job need to be tracked and reviewed. Every new effort should be tested and tracked against other methods of gaining work.

There is no such thing as a bad lead. Just slow, lazy contractors.

SALES, SALES, SALES

Sales is my favorite part of my business. By following the principles in this chapter, most home improvement contractors, either as individuals or as sales managers, can see quick improvement in their sales skills.

The first part of focusing on sales is to dispel any misconceptions about sales in general. The job of the owner or CEO of any business is to be the company's Chief Cheerleader and Salesperson.

If you are not excited about selling yourself and your company to your customers, then quit now and go get a job working for someone who is. Understanding that constantly improving sales skills is an "A" priority for his or her business is crucial to any contractor.

I am now officially done with the lecture.

SALES IS NOT A DIRTY WORD

As discussed earlier, many contractors are amazingly talented at their trade, and because of years of working for less-talented contractors, decided to start their own business.

They are, and should be, very proud of the quality of work that they perform. Their challenge in starting up a business, though, is in never having had the responsibility to talk to customers and to sell them on themselves and their company. They tend not to know how to be this Chief Cheerleader and Salesperson.

Then came along this book you're holding.

When many of us think of sales, we automatically think of used car salesmen like Robin Williams in *Cadillac Man*, or we think of the aluminum siding salesmen in *Tin Men* (which, by the way, are both very funny movies and worth watching).

In general, even the thought of sales can bring up an almost allergic reaction to some contractors.

Sales, though, is actually something that you do every day whether you notice it or not.

Sales is the simple art of communicating your ideas, thoughts, plans, and dreams to someone else. Sales is about finding out what someone else really wants and trying to see if what he or she wants is something that you can provide them.

We are selling daily and being sold to daily. Some of us are better at it than others, but it is utterly necessary for company survival. Not being capable of effectively sharing your ideas with others will have devastating consequences.

> *Sales is about finding out what someone else really wants and trying to see if what he or she wants is something that you can provide them.*

All around us are people who never go to the movies or restaurants they want to because someone else in their group convinced everyone that their movie or restaurant was better. People are wearing clothes and living in houses and towns that they really don't like purely because of their inability to speak up and share their desires.

People get passed over daily for promotions and end up reporting to a boss they hate because, even though their ideas were better, they could not sell them to the people in charge.

Relationships fail all the time because people are unable to take the time to listen and find out what the other person really wants in an effort to make that person happy.

These, too, are part of sales.

If there is one thing that you take from this book, it should be a thirst to learn more about sales and become the best communicator that you can possibly be. And if you don't, you will be doomed to a life of misery and you will never experience the joy of getting your own way while making everyone around you happy.

Is it really that magical?

Well, no, but there is some truth in the last paragraph. Selling is a part of your daily life, and as CEO and owner of your business, you need to be amazing at it!

> *If there is one thing that you take from this book, it should be a thirst to learn more about sales and become the best communicator that you can possibly be.*

THE BASICS

Call When You Say You Will

With cell phone plans being as inexpensive as they are, there is absolutely no excuse for not calling customers back when you say you will call.

I am sure you are extremely busy, but so are your customers—and they are waiting for your call. They are judging you on whether you do or do not do exactly what you say you will do.

The good news for many of us is that 20%-plus of home improvement contractors do not call when they say they will call. They are just "too busy" and "could not break away."

To these contractors, the rest of us would like to say, "Thank you for making it so easy for us to take jobs away from you!"

> *They are judging you on whether you do or do not do exactly what you say you will do.*

Call to Confirm

Calling to confirm an appointment is an often missed, but very important step. Always call the customer 15 minutes or so before your appointment just to let them know you are on your way and what time you will be arriving.

If you are going to be late, and it does happen, all tends to be forgiven with this phone call. If you show up late with no call, your chances of getting the job are a quarter of the chances of the contractor who does make the call.

Show Up

This sounds simple, but you will automatically beat 10–15% of your competition just by showing up.

Our industry has a horrible reputation for not showing up at all for set appointments. You've seen it. Hopefully, you haven't done it. Those who have just simply not shown up undoubtedly have amazing tales and excuses as to why they didn't keep the appointment. No matter the tale, there is absolutely no excuse for blowing off a customer.

If you are going to completely miss appointments, then do us all a favor and go do something else for a living.

You give the rest of us a bad name.

First Impressions

How you look when you show up at a client's home is crucial to their decision making.

You need to make sure, regardless of how your day has gone, that you are in a clean company logo shirt and clean pants with combed hair and breath that doesn't smell like the back of a dumpster.

Sound like basic stuff? Our industry has a reputation of not adhering to these simple and obvious sales principles.

You only get one chance at a first impression. Don't blow it!

Keep a clean shirt and pants on a hanger in your car or truck. Stash a bag of toiletries in the trunk. Whatever you need to look good should always be at arm's length.

Think of this initial meeting as a first date where you need to impress.

The Vehicle

Customers pay attention to the car you drive. Believe it or not, it matters what you drive to a customer's home.

Ideally, you arrive in a company vehicle with your company's logo on the side. (Sign companies can make a removable magnet of your logo for the side of any vehicle.)

The vehicle needs to be clean inside and out. Customers may glance inside your vehicle, see the pile of McDonald's® wrappers, and then make assumptions about you that cause them to run to other contractors.

Don't come driving up in a sports car or convertible. Customers associate these cars with people who are out to have fun and will assume that you will not take their job seriously.

Stay away, obviously, from the high-end vehicles such as Mercedes®, Porsche®, BMW®, etc. Even though you are doing well enough to own one, don't drive it to a potential customer's house. If you do, you'll soon not be able to afford the payments.

Customers don't want to pay for your excessive lifestyle.

Shoes Count

There are people who pay close attention to others' shoes. It definitely reflects on you. Keep a clean pair of shoes with you as well.

Homeowners don't want to see your work boots in their house. Their home is not a work site! Upon entering someone's home, always remove your shoes, or at a minimum, wear the little booties that go over shoes to protect the flooring.

The customer is watching and making mental notes of all these things.

More than 40% of your competitors overlook this shoe advice. Your customers do not!

If you are going to completely miss appointments, then do us all a favor and go do something else for a living.

You only get one chance at a first impression. Don't blow it!

Believe it or not, it matters what you drive to a customer's home.

Don't come driving up in a sports car or convertible.

Upon entering someone's home, always remove your shoes, or at a minimum, wear the little booties that go over shoes to protect the flooring.

Tattoos, Earrings, and Other Miscellaneous Stuff

There are those who enjoy cool tattoos, earrings, and flashy chains. A home improvement estimate visit, though, is no place to show off your bling.

Assume that any of this stuff that is not mainstream for a 55-year-old baby boomer is going to affect your chances of booking the job. Enjoy your weekends and personal time. Show off whatever you got. But, during the week, cover the tattoo and get rid of the bling.

Long Hair

Clean-shaven usually leaves a better impression when people are looking for someone to work on their home or property, and long isn't part of that look.

Don't like what you hear? Go ahead and take your chances.

Got Smokes?

Do not ever smell like smoke when you go into a customer's home.

People who don't smoke can be absolutely indignant about being around it. They will assume that you will smoke in their home while they are away (even though you try to convince them otherwise).

They will automatically pass you over for someone who does not smoke. There are studies that prove the point. Disagree at your own risk.

Do not ever smell like smoke when you go into a customer's home.

Smile All the Time

No one wants to do business with a negative and unhappy person.

They do not care about your personal life or how tough your business is right now.

Why would a homeowner want to do business with someone whose life is falling apart?

The simple answer is they don't. They will choose the contractor who is not a mess and who has a positive outlook on life.

Got a Funny Story?

If you have a good, clean, and funny story, use it to break the ice. Get the customer laughing and break down the barriers that automatically go up when you walk into a home.

If you don't have a funny story, then find one and practice it. Go ahead and use the same one all the time, just be careful not to tell the same story twice to the same customer.

If you have a good, clean, and funny story, use it to break the ice.

Ask Questions—Learn Something New

When I got started in sales, I absolutely loved in-home estimates. I would look forward to every single one of them.

I loved them because I knew that every day I was going to meet new people who knew things that I did not know.

I took full advantage of this opportunity to look around customers' homes and ask a bunch of questions. Asking what they did for a living would sometimes start an interesting conversation. Asking about a picture on the wall or the hobby being displayed out in the garage almost always would get a conversation going.

Every home I entered had something that would spark a conversation that I could then expand on by asking more questions. I gauged my success at the end of the day by what I learned at each stop—not by the deals I sold. (And, yes, I always sold a ton of deals!)

UNDERSTANDING YOUR CUSTOMERS

Customers are going to fall into a few main categories.

Customers are going to fall into a few main categories.

Hard Driving and Fast Talking

You know the type, not a lot of room for small talk here. They talk really fast, breaking down the wall that may be standing between the two of you can be very tough.

You need to move as fast as they do. They want to get quick and easy facts—so give them the information they want—quickly and easily.

This type of customer does not want their time wasted and only respects people who are the same as they are.

If you try to take too much time talking to them about something other than the purpose of your visit, they will then assume that you will drive them nuts when you are working on their house with your constant questions and chatting. They don't want to hire a "Chatty Cathy."

Get right down to business. Focus on how you understand exactly what it is that they want and how you will work to not disrupt their household. Convince them that you will be in stealth mode the entire time, just like a ninja, getting your work done and leaving without bothering anyone.

This type of customer does not want their time wasted and only respects people who are the same as they are.

The Technician

Technicians are the accountants, engineers, doctors, and people like my brother-in-law who actually enjoy curling up with a good owner's manual (no joke).

These guys want the details. They want *all* the details. They want to know the exact make up of every product you will be using and why you have chosen them. They want to know about your crew and who will be working on the job. They are interested in your hiring procedures and the details about you and your company.

They are not necessarily on a time frame and could spend most of the day with you (even for a $500 job). They are offended if you try to rush them, and they appreciate follow up emails with more information like product spec sheets.

The most important thing to remember with this type of individual is never to make up anything. They will know the minute you are lying. They have an extra gene that allows them to determine who is telling the truth and who is not. Ignore this one and you'll get caught.

If you don't know the answer, simply write it down and tell them you will find out for them. Answer all of their questions truthfully and honestly to the best of your ability and you will earn their trust.

If you don't know the answer, simply write it down and tell them you will find out for them. Answer all of their questions truthfully and honestly to the best of your ability and you will earn their trust.

Friend of the World

These people are the best. They want to talk and tell stories, make you lunch, and then have a beer with your when the job is done.

They will tell all their neighbors and friends what a great job you did in ways that will keep you awash in work for months—IF you take the time to talk, laugh, smile, and share stories with them.

If you rush them or appear not to care about what they are trying to share with you, then they will not trust you. If they don't like and trust you, then they will not do business with you.

Salespeople

Salespeople are so special that they deserve a category of their own.

I love selling to salespeople for one reason and one reason only.

Salespeople love the game. They love to be sold to.

They are preconditioned after years of sales training to buy! They just want to be certain that you gave them your best presentation and made sure they had fun while you did it.

One of my favorite stories happened when I was out training a new salesperson and we were at the point of "closing the deal." The young salesperson I was training asked for the job once and the "customer salesman" said, "No, I'll have to think about it." The rookie got up, said "Thank you," and excused himself. The customer and I were still sitting at the table.

The customer turned to me and asked, "Is he really going to give up that easily?" I said, "He may be, but I'm not." After 45 minutes of going back and forth, the customer signed the deal, looked at me and said, "Thanks. That was a lot of fun."

He then turned to the rookie and said, "I hope you learned something today." Salespeople all love the game and just want to be sold to.

Salespeople are so special that they deserve a category of their own.

Salespeople all love the game and just want to be sold to.

Got Nothing

The most challenging type of person to deal with is the one who gives you absolutely nothing to work with.

They are a blank slate—emotionless and disinterested. (The only thing that could possibly be worse than trying to sell to one of these folks is to be in the same family with one!)

The only thing you can do with the small part of the population that is completely uninterested in human communications or relationships is to be as professional as possible, stick to your presentation, and hope for the best.

When you get the signals that they are not interested in talking about their families and jobs or listening to your amazingly funny stories, then back off immediately and stick to the facts.

The most challenging type of person to deal with is the one who gives you absolutely nothing to work with.

TWO EARS AND ONE MOUTH—USE THEM PROPORTIONATELY

The magical key to selling jobs in home improvement is to do one basic thing.

Listen.

Most contractors will show up at the customers' home and proceed to talk their ears off about why they are superior. Even worse is showing up and, whether because of talking too much or not, failing to listen to the customer!

If you listen close enough and ask the right questions, customers will tell you exactly what is going to be important in their buying decisions.

WHY PEOPLE BUY—BECAUSE YOU ARE COMPETENT

Customers want to know that you are good at what you do.

They want to see pictures and reference letters from people for whom you have done work in the past. They want phone numbers to call (even if they don't intend to call them).

They need to know that you fully understand your trade and that you are a professional who has chosen to do this as your life's work.

If they believe this to be true, you are closer to booking the job.

Trust

If you lie, they will catch you.

If you try to make things up or look the least bit shady, they will pick up on it and not give you the job.

If you don't show up when you say you will, or forget to call when you said you would, they will not give you the job.

Safety

A large number of home improvement contractors forget that they work in people's homes. Their home is their castle. It is something that they actually have control over and, more important, it is where their families live.

Everybody will do everything possible to make sure that their homes are safe for their families, including being very careful about who they actually allow into their sanctuaries.

Homeowners base a significant portion of their decision to hire someone on the degree to which they feel like the contractor is safe to have in their homes.

They are highly in tune with these protective feelings. If they fear even in the least for their safety, if for some reason they don't trust you, or if something in their gut is uneasy, they will toss you away like day-old donuts.

Take this warning seriously: if you are not someone that they would like to have over for lunch with the kids, then they will find someone who is, and that person will get the job.

THE PROCESS

All the best sales presentations have a process. They are done the same way each time. A good sales presentation is just like a play. Think of it as a Broadway show. In our Sales Show, there are Acts 1 through 11.

Act 1: The First Phone Call

The first phone call is really the very first impression you are going to be leaving your customer. They will either leave the phone call looking forward to meeting you or questioning why they called you instead of someone else.

If you listen close enough and ask the right questions, customers will tell you exactly what is going to be important in their buying decisions.

Customers want to know that you are good at what you do.

If you lie, they will catch you.

A large number of home improvement contractors forget that they work in people's homes. Their home is their castle.

Take this warning seriously: if you are not someone that they would like to have over for lunch with the kids, then they will find someone who is, and that person will get the job.

The first phone call is really the very first impression you are going to be leaving your customer.

Take the time to ask what the customer is looking to have done. Get as much detail as you can.

Obtain the following information:

- What is the scope of the project?
- What experiences have they had in the past with your trade?
- Do they have any special concerns?
- What is the project frame?
- What is their budget?

Phone tips:

- Have your 60-second commercial down cold. Practice makes perfect, so practice!
- Do your best to make sure everyone who has an interest in the project will be home when you come over to do the estimate.
- Smile while you're on the phone—your customers can tell the difference.
- Speak clearly.
- Make sure you are in a place with no background noise.
- Write everything down, including the names of their dogs and kids—and bring the notes to the estimate.

Have your 60-second commercial down cold.

Write everything down, including the names of their dogs and kids— and bring the notes to the estimate

Act 2: The Front Door

This is the part where you need to go back and reread the section on first impressions. Success or failure is dependent on these first impressions.

Act 3: Time to Chat

Act 3 is the part of the show where you learn about the customer.

Take some time to chat casually and build some rapport. This part of the show should not be rushed, unless, of course, the customer wants to rush it.

In general, the longer you are able to continue building your relationship with the customer, the better chance you have of closing the job.

In general, the longer you are able to continue building your relationship with the customer, the better chance you have of closing the job.

Act 4: The Work

This is the part of the show where you find out what the customer is looking to have done.

This is why you are there. Listen carefully and ask a lot of questions. With the notes from the first call, specifically ask to see the areas that they told you about.

- What work is to be done?
- What is the scope of the work?
- Why do they want to have it done?
- What areas are they concerned about?
- What are their expectations for time frame and quality of work?

Take lots of notes. It is impossible for you to remember all the details of what your customers tell you. These notes will come in handy as you prepare your estimate. It also demonstrates to the customer how detail oriented you are.

Act 5: Your Time to Shine

Act 5 is where you get to show off your stuff. All those years of hard work and knowledge of your trade are about to shine through.

Act 5 is all about you as a professional home improvement contractor showing how you are going to solve your customer's challenges.

This is where you talk about the way in which you will tackle the job, the products that you will use, and the extensive experience that you have with this type of project.

This is where you show the customer things that they may have not thought about yet. Talk about the specifics that they as a "lay" person may not be aware of.

The most important thing about Act 5 is to realize that it is Act 5 not Act 1.

You need to go through all the first acts in their entirety and not just rush to Act 5.

If you walk into a play right in the middle of it, you will not understand or appreciate it. You need to see it from the beginning to be able to see how each act built upon the other acts leading up to the act you just walked into.

Act 6: Your Reference Book

Now that you fully understand what the customer is looking to have done, you can walk back over to the kitchen table and show them your reference book.

Take the time to show them your contractor's license (if necessary) and your insurance information. Pictures of your family and crew are always a great touch. Include trade associations that you belong to and, of course, include pictures and reference letters from local customers.

The last thing from the book you should show your potential customer is a section on the products and materials that you use, focusing on the fact that you only use the highest quality materials.

Act 7: The Measure

This is the part where you leave the customer to review your references and talk among themselves, while you review the project, take measurements, and then go back to your car or truck to write up the bid.

Write up the bid?

Yes, you will never have a better opportunity to close this deal than right now.

If you have taken the time to go through all the first six acts and not rushed the process, then, as you are taking your time to get your measurements done, your customers are already making their decision.

They have already decided if they like and trust you. If they do like and trust you, then it is simply a matter of how well they liked your professional presentation and whether your price is in the general ballpark of their budget.

Take lots of notes. It is impossible for you to remember all the details of what your customers tell you.

The most important thing about Act 5 is to realize that it is Act 5 not Act 1.

The last thing from the book you should show your potential customer is a section on the products and materials that you use, focusing on the fact that you only use the highest quality materials.

Tips for writing up the bid:

- Print, don't write. If your handwriting/printing is poor, fix it. People view a sloppy contract as an example of the work you will do.

- Use a proper three-copy contract with your logo and information on it.

- Make sure you have spaces for the customer's home phone, cell, and email.

- Write on the contract specific areas of concern the customer had.

- Be extremely specific about the work being completed.

- Remember, you only get one shot at this, so don't mess it up—literally.

- Don't forget to include a very clear and legible price with all options listed in detail.

Taking your measurements home and getting back in touch later is an absolute no-no. Only the huge jobs (full remodels, etc.) require a second visit.

Never—ever—mail, fax, or email a bid. You need to be in person when a decision is being made. If you are making a second visit, start again with Act 1. You may need to go through those first acts a little faster this time, but your customer may have forgotten why they like you. You need to remind them.

Act 8: The Big Finish—The Climax of the Show

This is it; this is why you actually came to the customer's house.

This is now your opportunity to close the deal, to walk away with a signed contract and a job for you and/or your crew to start.

This is the moment where your hands get a little sweaty and you start mentally questioning your numbers.

Are you too high? Too low?

Now is the time to trust your estimating procedures and ignore the price.

The price is the price is the price. It is only one small part of what makes a consumer decide who to let into their home.

Do you really think that to save a few dollars the customer will hire the ex-con who has been asking questions about the house alarm system?

They will hire the person who they trust the most.

Is that person you?

Steps to the Close

- Get the customers to join you at the kitchen table, making sure anyone who is going to have a say in the project is sitting down with you.

- Sit down and place the contract in front of them on the table.

- Have a pen in your hand for the signature (I always had my lucky pen—every salesperson is a little superstitious).

- Get the price out of the way. "After going through your project in detail, I have come up with a cost of X."

- Now move on. "What I would like to do now is go through the estimate with you and make sure that I did not miss anything and that I am providing you with everything you asked for."

If your handwriting/printing is poor, fix it. People view a sloppy contract as an example of the work you will do.

Taking your measurements home and getting back in touch later is an absolute no-no.

Now is the time to trust your estimating procedures and ignore the price.

Do you really think that to save a few dollars the customer will hire the ex-con who has been asking questions about the house alarm system?

- Go through the estimate, making sure to restate all of their areas of concern.
- Double check that you have their correct phone numbers and email address (just in case they don't sign the job on the spot).

Ask Closing Questions

- Is this the job you were looking for?
- Is it OK to have a small space in your garage for the equipment?
- Is this the exact color that you were looking for?
- I am going to spend a little extra time on this, because I know you were concerned about it. Is that OK with you?

Going for the Close

"I am confident that I will give you exactly the job you are looking for. Do you have any questions?"

Remember after you ask a question like this, you need to SHUT UP! Sales is all about knowing when to be quiet.

Sales is like a game of tennis. Once you have served the ball and it has gone over the net, what do you do? You do nothing more than get into position and wait for the ball to come back to you. You are done. You don't sit there and keep swinging the racket like a madman. You get into position and wait for your turn. Once you have asked a question, you need to sit there quietly and wait for an answer. It does not matter how long it takes. (If your customer is a salesperson, they will mess with you by sitting there and not saying anything for as long as they can, just to see if you blow it!)

This is the art of the close!

> *Once you have asked a question, you need to sit there quietly and wait for an answer.*

Back to Our Show

Once you have answered the customers' questions or found out that they have none, your next step is to move in for the kill.

"Well then, the only thing left is for me to get your signature down here at the bottom of the agreement."

Now put the pen down and shut up!

You don't need to do anything else. The customer knows that it is now their turn. They completely understand that you are expecting an answer from them.

They understand that you want them to pick up the pen and sign the agreement. You don't need to jump in and save them from this sometimes awkward moment. Just like in a game of tennis, the ball has been hit into their side of the court at 70 miles an hour and is coming straight at them.

They will react!

> *Objections are opportunities to educate your customer on things that they may be misinformed about or just need to be reminded of.*

Act 9: Objections

Objections are opportunities to educate your customers on things that they may be misinformed about or just need to be reminded of.

Classic objections that you may encounter:

- Not enough trust or confidence in your company.
- Not enough trust or confidence in you.
- Price—Does your price fit their budget?
- Price again—Are they realistic about what they want to have done?
- Decision process—They cannot make the decision on their own.
- Lack of education—They want to get other bids.

What are the typical objections that you come across, and how do you handle them?

Act 10: Keep Closing

This is a long play. Get used to it! This whole show can last an hour and a half to two hours.

Close at least three times. (The reality is that you are supposed to close five times, but let's start easy.)

After each objection that you have handled successfully, you need to go back and once again ask the tough closing question. If you want the job, you are going to need to ask for it more than once.

"No"—in Sales—Does not Mean "No"

Most customers are preprogrammed to tell you that they are going to get three estimates and that they will not be making a decision today.

They will typically say "No" the first couple of times that you ask them for the job simply because that is what they believe they are supposed to do.

If they like you, they are simply waiting for you to finish getting them comfortable before they say "Yes."

They want to say "Yes." It is time-consuming and a hassle to get other estimates. They already trust and like you. Keep asking for the job (just not the same way each time) and you'll get it.

Act 11: Now Back for Another Command Performance

Yes, there are times when even the best of us are not able to convince our customers to sign up with us on the spot. Even though you should be able to close over half of your proposals on the spot (assuming no plans need to be drawn, etc.), there will be times when you will need to take the time to enter the final act of this play.

It is shocking the number of contractors who never call the customer back after they do an estimate.

These contractors are either so hurt by the fact that the customer did not choose them right there on the spot, or they are under some misguided notion that the customer will, of course, call them back when they are ready to have the work done.

Closing jobs after you leave the customer's home requires a very specific step-by-step process.

Close at least three times. (The reality is that you are supposed to close five times, but let's start easy.)

It is shocking the number of contractors who never call the customer back after they do an estimate.

Closing jobs after you leave the customer's home requires a very specific step-by-step process.

Step One

Find out why they are not ready to agree today.

- Is it the financing?
- Do they want more estimates?
- Did you fail to win them over with your amazing 11-act show?
- Ask as many questions as needed (without being too pushy) in order to find out exactly why they feel the need to wait.

Step Two

Set up a specific time to call them back in the next 72 hours when they will be able to make a decision.

Step Three

Confirm cell and home numbers, and email address to make sure you can get in touch.

Step Four

Send them a thank you card—not a thank you email—thanking them for their time and letting them know how much you are looking forward to doing their job. (Sendoutcards. com™ will make this easy for you.)

Step Five

Call back when you say you are going to.

Step Six

Look for information regarding their job and send them email updates on products and maybe another reference letter. Send them a couple of emails with these updates.

Step Seven

If they have not made their decision when you call back, set up another specific time to call back.

Step Eight through?

Repeat Step 7 until you have booked the job or they tell you to leave them alone.

Extra Credit

If you are in the neighborhood, drop by and say hello with a dozen donuts.

Conclusion to Our Play

This concludes our 11-act play. I hope you have enjoyed it and please come again!

REHEARSAL

Actors rehearse their lines over and over again until the lines become burned into their memories and they can perform without thinking about what to say.

Golfers, martial artists, and baseball players practice hour after hour on the same strokes, moves, and swings. Practice causes the movement to get burned into muscle memory, so that they can perform the action without thinking.

In Malcolm Gladwell's book *Outliers*, he talks about how the main difference between musicians who were great, good, and okay was the number of hours they practiced over their lifetime. The top musicians had put in close to 10,000 hours, the good ones 6,000 hours, and the okay ones 3,000 hours or so.

The difference between okay and great was all about the practice time and not so much about a God-given talent.

You can relate to this in your own trade. There is a definite difference in someone who has logged 10,000 or 20,000-plus hours in a trade versus the new guy who is only in year two.

Experience Teaches Us Many Things

Salesmanship is no different than music and athletics. You and your salespeople need to practice.

Go through your presentation until it is burned into your mental memory and you start saying it in your sleep.

Know and understand the common objections in your trade, write out the best answers, and then practice them over and over again with friends and family. Rehearsal is required to get you past "okay" and "good" and make you "great" at sales.

THE TOOLS

Product Information and Samples

What products will you be using in or on the customer's home? It is not just about you and how you show up; it is also about the quality of materials that you plan on using in a customer's home.

Props Make All the Difference in a High-End, In-Home Presentation

Props, props, and more props. All the world is a stage.

Every one of you has a few products that lend themselves to a little extra special attention and that are something that customers would be intrigued to touch, hold, or just watch a video of.

Window Replacement

Years ago I started a window replacement business in Illinois. It did not work out as planned (which could be a book in itself!). I went to two days of training with our manufacturer in order to learn how to sell its product. I must say, I was very impressed.

The difference between okay and great was all about the practice time and not so much about a God-given talent.

Salesmanship is no different than music and athletics. You and your salespeople need to practice.

Go through your presentation until it is burned into your mental memory and you start saying it in your sleep.

Every one of you has a few products that lend themselves to a little extra special attention and that are something that customers would be intrigued to touch, hold, or just watch a video of.

I saw windows that reflected heat, and were easy to clean inside and out, and some windows that had so much tint they could turn your house into a rapper's limo.

My favorite demonstration of all was one where they pulled back a bowling ball on a rope and let it go, allowing it to swing right into this glass window without breaking the glass!

I loved this stuff. The greatest thing of all was that for only $200 per salesperson, the manufacturer sold us these nice kits packaged inside of a small suitcase, ready for my salespeople to take out and do the same demonstration for our customers (the bowling ball, unfortunately, was replaced by a hammer—yes, I was disappointed).

Those customer presentations were over the top!

Painters

There are some amazing products out there. One is fireproof paint. While no one wants to watch a demo of paint drying, who wouldn't be riveted while watching a video of a room coated in this fireproof paint not burning after a Christmas tree, sitting in the middle of the room, is set on fire?

Now hand the customer a lighter and a stick covered with a coat of this amazing paint and ask him or her to try and light the stick on fire (preferably outdoors or over a fireplace).

Plumbers

PEX is an amazing product that is used instead of copper plumbing.

Hand your customers the simple blue and red tubing and let them play with the easy valve system. Show them how they can now control the hot and cold water from their attic or basement locations.

This is an especially cool product for new construction.

If you sell reverse osmosis (RO) water systems, the prop world is your oyster.

There are great props that show just how bad the water is that comes straight out of the tap (even out of store-bought bottled water).

Electricians

There are almost too many ideas here to write down. Here is a simple one. If you want to become a trusted advisor, bring a phantom power meter to your next estimate.

Show your customers just how much electricity all of their TVs use when they are supposedly off. After such a presentation, the easiest sale you'll ever make is pitching those power switches that you have in your truck.

> If you want to become a trusted advisor, bring a phantom power meter to your next estimate.

The Wow Effect

Props are only good if they have a *"wow effect."*

Customers need to be able to look, touch, and feel your prop and say, "Wow, I've never seen anything like this before!"

Answer these questions about the props you use to sell your business:

- What props do you currently bring on a presentation?
- What is the wow effect from these props?
- What could you use to get that wow effect?

> Props are only good if they you have a "wow effect."

Where to Get the "Wow Effect"—Suppliers

Products exist that have the desired wow effect.

Go to your trade organization shows and walk the floors. Go have lunch with reps from different suppliers. These guys have teams of people who have spent years building kits for in-home demonstrations and are sitting around wondering why no one wants to use them.

What does your customer need beyond what you are currently doing for them?

Don't Over-Prop

A word to the wise. You are better off having two or three really cool things that will deliver the wow effect, instead of eight things that will put your customers to sleep.

Over-propping can kill a sale. How many props should you use? It depends on the customer's personality type as mentioned earlier. This is another reason why it is important to read your customers and understand what they want to see and hear.

UP-SELLING

Up-selling is the act of discovering what more a customer needs and making sure you are the one to deliver it.

Every job you go on has the opportunity to become a larger and more profitable job for you and a better experience for the customer. The key to making this happen is to genuinely evaluate what your customers could need outside the average job.

What does your customer need beyond what you are currently doing for them?

The Basic Up-Sells

The basic up-sells are the things that you are already doing (or should be doing).

For the electrician, they can be as simple as identifying faulty wiring or bad light sockets that you happen to come across while you are at the home.

For the plumber, they can be checking sinks and fixtures and offering to replace or fix those as well.

For the handyman, the list is endless. Look around. Take notice of what else needs to be done?

The painter can easily add an extra room or hallway or add an accent color.

The landscape contractor can quickly do a check of the sprinkler system and offer to repair broken sprinkler heads.

These are the simple things that everyone should do with every customer they have. The average job size goes up and your customer has a much better experience because you were able to find and fix things that they may not have seen.

Everyone wins.

What are the five basic things that you should be looking for or doing on any job in order to add more money to a contract and leave your customers with a better experience?

The basic up-sells are the things that you are already doing (or should be doing).

1. _____

2. _____

3. _____

4. _____

5. _____

The Quality Up-Sell

Everyone should have a good/better/best strategy. Every customer is different and every customer decides to buy or not to buy based on her own set of criteria—not yours.

If you have not been offering a good/better/best option, then you have lost work to your competitors. Your competitors understood that this particular customer was not just looking for the cheapest price, they were looking for the highest quality workmanship and products.

In almost any area of contracting, there is a way to offer a better quality result by either spending more time or using different techniques or better materials. Let's take a few minutes to put down in writing some good, better, and best options related to your trade.

The Good

Think about your typical job.

Materials: What is the typical grade of materials that you would use and how much of it would you use?

1. _____
2. _____
3. _____
4. _____
5. _____

Time: How much time does it take to complete the project from start to finish?

_____

Pricing: What is the complete price of the project, including all labor, materials, overhead, and profit?

1. _____
2. _____
3. _____
4. _____
5. _____

Better

For this option you need to take the same job and think about what would make it a better quality and longer lasting finished product.

Materials: What materials exist that could make this job better than the "good" job? What could be done from a workmanship standpoint to add value to the job?

Make a list of possible upgrades for the "better" job, including costs.

1. _____
2. _____
3. _____
4. _____
5. _____

Pricing: Add up the cost of doing this "better" job and reprice the contract taking into account the higher quality materials and the additional time it takes to complete the revised project.

Obviously, this should be more expensive than the "good" job.

1. _____
2. _____
3. _____
4. _____
5. _____

Best

Here is the fun one. This is the same good job, but with only the highest quality materials that you can get your hands on, as well as the proper amount of additional time taken to make sure the job is a work of art.

Materials: Make a list of the best quality products that can be used for this project.

1. _____
2. _____
3. _____
4. _____
5. _____

Pricing: Now add up the pricing for the "best" job, including the new materials costs, labor, overhead, and profit. This should now be an even higher price than the "better" option.

1. _____
2. _____
3. _____
4. _____
5. _____

Your Customers Want Options

Right about now many of you are saying, "Look, Jay, my customers will not pay for all these added things . . . all they ever care about is price."

Customers are not all driven by price, even though they will tell you that is the only thing that matters.

If price was the only thing that mattered then we would all drive the same type of car, live in the same style of house, and wear the same kind of clothes (soccer moms excluded, of course).

People make their decisions based on their different "hot buttons."

By only giving your customers the option you feel they asked for (usually the "good" option), you are not allowing yourself the opportunity to sell customers something more that they haven't yet considered.

You are also leaving bigger jobs with larger profits on the table for your competitors to grab.

Customers are not all driven by price, even though they will tell you that is the only thing that matters.

People make their decisions based on their different "hot buttons."

Green Options

Green is in! If you don't believe that, just turn on the TV and spend a couple of nights watching the Planet Green™ channel.

You will be amazed at all the different "green" building materials and options that exist. Studies show that customers will pay upward of 25% more for a green option.

The first step is to actually figure out what "green" means in your particular industry. Your trade association magazines and Websites are a great source of information for what your trade is doing to be green.

Why Go Green?   Just in case you have been living under a rock and have not been paying attention to the news for the last 10 years, you need enlightenment: Everyone is going green!

We, as Americans, have finally woken up to the perilous environment and all the harm we have been doing to it.

Going green helps save the environment for our kids and makes us feel better about our own contributions to the world. It also allows us bragging rights around the water cooler. Believe it or not, it is cool to go green!

What are the top green building materials in your industry that you can offer as an option?

Just in case you have been living under a rock and have not been paying attention to the news for the last 10 years, you need enlightenment: Everyone is going green!

1. _____
2. _____
3. _____
4. _____
5. _____

Be an Energy Star!

Another great missed opportunity for contractors is showing customers how they can save money and help the environment, all the while spending more money with you.

Every industry has energy- or water-saving options.

Most customers have heard of the options, but don't have a great understanding of how they work or where to get them. You can educate them on their different options, and they can buy them from you. The opportunities are amazing.

Here are a couple of quick examples of options that exist.

Plumbers   Amazing opportunities exist for every plumber. Almost everyone would love to see their water bill go down.

The options are endless. Offer your customers dual flush toilets, low-flow shower heads, tankless water heaters, and separate under-the-sink instant water heaters.

Electricians   Most people have heard of compact fluorescent lighting, but the vast majority of households have not replaced all of the incandescent bulbs with these CFLs.

Many people have it on their to do list, but don't have the time to purchase and install them or understand what a great savings they could see on their electrical bill.

What if you always carried a supply of CFLs and offered to install them?

Simple dimmer switches save money, as do programmable thermostats, etc. The list goes on.

Take the time to offer a home energy audit and show your existing customers how they can save money on their electrical bills.

Take the time to offer up a home energy audit and show your existing customers how they can save money on their electrical bills.

Landscapers   The majority of people's water expenses come from irrigation.

Offer green alternatives, replacing older inefficient sprinkler heads and replacing or repairing drip irrigation or sprinkler systems. Replacing grass areas with native drought-resistant plant areas can save your customers huge amounts of money on their water bills.

Your Business   In what areas can you save your customers money and make the job more profitable for you?

1. _____
2. _____
3. _____
4. _____
5. _____

Be a Trusted Advisor   Being an energy star is all about taking the time to find out how you can use your knowledge and trade to save your customers money and, in turn, make more money for yourself. The side benefits are obvious. Customers will rave about you. Everyone loves someone who teaches and helps them.

You can also transcend being just a contractor and become a trusted advisor. A trusted advisor is someone who a customer will repeatedly go to for advice, not to just fix things.

Of course, you will also feel pretty good about your contributions to the environment.

The Fun Stuff Most People Don't Think About

What is new and cutting edge in your industry that you could offer to your customers that most of your competitors don't offer yet as an option?

Every day, new products and materials are created that change the landscape of contracting.

- **Painting:** Rustoleum® has a magnetic paint that can transform any wall in any room into a "refrigerator door" ready to tack up the kids' artwork. There is also a product that turns a regular wall into a chalkboard. Imagine the fun kids could have with that!

- **Electrical:** Powersave® has a product called the Powersave 1200® that, when added to an electrical box, can cut down your energy use by 10–15%. (It works! I have one.)

- **Landscapers:** How about offering a satellite-controlled watering system that takes control of customers' outdoor irrigation systems so that they don't need to adjust for warm or cloudy days?

- **Your Business:** What unique products exist in your industry that could be up-sells for your customers?

1. _____

2. _____

3. _____

4. _____

5. _____

Up-Selling Summary

By offering your customers choices and educating them on options that they never knew existed, you have the opportunity to change from being just another contractor to their trusted advisor. Trusted advisors have a different level of respect.

Trusted advisors have a different level of respect.

This added respect leads to increased average job size and profit. You will be able to leave your customer in a much better place than if you had just gone in and done the lowest-cost job possible.

Your most-profitable and busiest competitors are probably doing this already! Now it's your turn.

SALES MANAGEMENT

You cannot grow your business beyond a certain point without having people out there selling for you.

The challenges related to hiring, training, and managing salespeople in the home improvement world are a company's single biggest roadblock.

The two limitations to growing a home improvement business are:

1. Marketing (go back and reread Chapter 2)

2. Sales

THINK YOU ARE READY TO HAVE A SALES TEAM?

Before you hire salespeople, you need to ensure that you are ready to have them on board. Also, not wanting to do the sales yourself is not a good enough reason to hire salespeople.

No salesperson should be able to outsell the owner of the company. If you are not confident that you can outsell any person that you hire, then don't hire anyone!

Take the time to go back to your sales process and become great at it. If you are horrible at sales and you go out and hire and train salespeople, how good will they possibly ever be?

Do You Have a Strong Lead Generation Program?

Notice I did not ask you if you understood your branding message or if you had a solid marketing program.

I specifically asked if you had a strong lead generation program. If you are unclear on what I mean here, go back and reread the lead generation section in Chapter 4.

If you have a strong lead generation program, then you will know that you can generate a certain amount of leads weekly at a price point that fits into your marketing model.

In other words, you know your numbers cold. For example, you know that it costs you $40 per lead and you personally can close one out of every four leads. This results in a marketing cost per job of $160.

No salesperson should be able to outsell the owner of the company.

If you are horrible at sales and you go out and hire and train salespeople, how good will they possibly ever be?

You also know that your salespeople will not be as good as you are out of the gate because of the "owner effect." You know that your business model still works when they close only one out of six leads.

You are confident that you are currently generating more leads than you can get to and that the trend will continue.

Production Capacity

Is your production team ready for the extra work that more sales will generate?

Will you have to hire extra crews? Will you need more trucks and extra equipment?

Cash

Expanding your business and hiring salespeople can be a very expensive endeavor. The cost of hiring and training salespeople, including salary, expenses and the leads they will burn while getting trained, can add up.

It is not uncommon for new salespeople to not start paying for themselves for several months.

Expanding your business and hiring salespeople can be a very expensive endeavor.

Are You Going to be Providing All the Leads?

If you are not going to be supplying all the leads (which isn't something you need to do—this topic alone could be another complete book), what are the activities that your salespeople will be doing to source more business? How do you know that these methods will work?

Have you personally done these activities so that you know the results that you should expect?

HIRING OF SALESPEOPLE

The best phrase I can think of to describe the hiring of a salesperson is *"caveat emptor"* (or translated, "buyer beware").

I am always suspicious of salespeople when I interview them. They are salespeople. It is difficult, sometimes, to tell if what they are saying is actually the truth or not. They are prone to exaggeration and many will say almost anything to get the deal or the job.

The best phrase I can think of to describe the hiring of a salesperson is "caveat emptor."

Experience is Great, But . . .

I have been in sales my entire adult life. I consider myself much better at marketing and sales than I do at the trade I chose.

Honestly, back to the book *Outliers*, I probably have well over 20,000 hours of experience in sales and management, but only a few hundred hours in my trade of painting.

Trust me . . . you don't want to buy any book that I write about "painting"—I am in no way qualified to author such a book.

I do, however, know salespeople. After more than 20 years of going through the process of hiring salespeople, I am convinced that a salesperson is one of the toughest hires out there. It is still tough for me. After all these years, I still consider myself fortunate if I am right 40% of the time. (Truth be told, I am probably wrong 75% of the time!)

Things to Keep in Mind

Why are they looking for work?

Assuming that you did not approach them directly and that they are not already gainfully employed, why are they looking for work?

Every employer in every industry understands just how valuable a top salesperson can be. One that truly delivers without an attitude is an amazing find. Top salespeople are never let go, no matter how tough the economy is, without an underlying reason. You need to find out what that reason was.

Top salespeople are never let go, no matter how tough the economy is, without an underlying reason.

Cultural Fit

How will they fit in with your organization? A top salesperson that no one in your organization wants to talk to or even go near is not going to work out, no matter how great they may be.

If you have a very team-oriented environment and your new salesperson is a prima donna or lone wolf, then trouble is right around the corner.

If you have an organization where employees go home at 5:00 and never see each other after work, and your new salesperson comes from companies where employees always hung out together after work for drinks, etc., then you will ultimately have a problem.

They're not going to enjoy the environment, will not be happy, and will start looking for a new place to work.

If you have a very team-oriented environment and your new salesperson is a prima donna or lone wolf, then trouble is right around the corner.

Will You Enjoy This Person?

Unless your organization is large enough to have a sales manager, you are going to be spending a ton of time with your salesperson.

Daily contact is essential for the "proper care and feeding" of a salesperson. I have a hard and fast rule about hiring salespeople. I have to like them as a person and I have to truly want to spend time with them. If not, then this hire is doomed to failure.

What Do They Believe?

What salespeople truly believe is more important than the actual truth.

There is no match if potential hires believe that in order to get a sale that they must offer the lowest price no matter what (and if you are not designed to be low price).

If their vision of follow up and customer care does not match yours, you will have a problem.

What salespeople truly believe is more important than the actual truth.

If they believe that the world should revolve around the sales department and you believe everyone is equally important, then trouble will brew.

If they believe that once the deal is sold, they're done and want their commission paid, but you want them to service the job while in progress, then they're not a good fit.

If they believe that it is your job to supply all the leads and you feel otherwise, then they're not going to be productive.

You need to make sure that you are 100% aligned in all of these areas. This is the same as getting married and not having discussed with your spouse how she feels about having children. Not being aligned on any of these points is a recipe for disaster!

Home Improvement Sales Is a Different Animal

Just because someone was successful in sales in one industry does not necessarily mean that the skill will transfer to the home improvement industry.

A skilled plumber is not necessarily a good electrician or general handyman.

A successful software salesperson does not always make a great home improvement salesperson.

Take, for example, someone coming from an inbound sales organization. They are used to the phone ringing with customers excited to talk to them about ordering a product. Their sales may be impressive, but their background and skills are not the same as having what it takes to go into a customer's home and put on an 11-act show!

How about someone who has had success in business-to-business sales? It is still sales and they were successful, but what hours did they work? Monday to Friday, 8:00 a.m. to 5:00 p.m. is not generally the best time to sell to home-owners. The best time, of course, is when both homeowners are home, which is almost always after 5:00 p.m. during the week and during the day on weekends.

Saturdays and Sundays are the best sales days in the home improvement world. If the person you are hiring is not used to working evenings and weekends, then you don't have a fit.

Are They Really Used to Selling or Are They Truly Marketers?

For years, I have had an argument with a friend in the pharmaceutical sales industry. He works in the pharmaceutical business in either a Director of Sales or VP of Sales capacity.

He would tell me about their salespeople. I would, in turn, argue that if somebody is not closing a sale, he is not a salesperson, but a marketer. A salesperson, in my mind, is someone who actually is required to "close the deal." They are required to bring home a signed contract in order to get paid. If they don't do this, they are not salespeople, they are marketers, simply promoting the product. (The great thing about writing your own book is that you win all the arguments!)

Home improvement sales require someone to close a deal and bring home a signed contract.

Just because someone was successful in sales in one industry does not necessarily mean that the skill will transfer to the home improvement industry.

If the person you are hiring is not used to working evenings and weekends, then you don't have a fit.

Home improvement sales require someone to close a deal and bring home a signed contract.

Trade Knowledge and Estimating

Whoever you hire is going to need a basic knowledge of the trade and be able to follow the estimating procedures that you set up.

Some people are just not handy or coordinated in any way whatsoever.

We all have friends who really have no idea how to change a light bulb (seriously). These guys, while great salespeople in other industries, do not belong anywhere near your customers' homes.

Estimating Procedures—You Gotta Have 'Em!

If you want to grow and expand your business, you absolutely need to have an estimating procedure. Whether it is from a trade manual or one that you created yourself, you cannot expect salespeople to have your level of "gut" understanding of how long a project will take.

Always Hire More than One

You will never be finished hiring salespeople. It is a huge investment and one that has the capability of taking your company to the next level.

It is also a project that never ends.

Chances are you only have a 25% chance or so that the person or people you hire will work out. You need to be constantly on the lookout for more applicants.

Never hire just one salesperson. It is much easier to manage a group of salespeople who now are forced to compete with each other than it is to manage just one who will start to believe after a few sales that you need him more than he needs you.

THE PROPER CARE AND FEEDING—ASSEMBLY REQUIRED

Salespeople need to be trained not only in your industry, but in how your company operates.

There are simply no out-of-the box, ready-to-work salespeople. New salespeople require more than just "some assembly." "Full assembly required" is your task (and batteries are not always included). Every company and culture is unique and your new salespeople need to be brought up to speed on all the nuances of your organization.

Ride Alongs

You need to be willing to commit to going out on the first 10 calls with new salespeople in order to make sure they are effectively communicating your company's message.

Are they performing the 11-act play exactly the way you wrote the script, or are they putting in their own additions that contradict it?

After making sure that they are doing things the way you want them to, you still need to be prepared to go out in the field with them once a week for the first year and once every two weeks until you retire.

If you want to grow and expand your business, you absolutely need to have an estimating procedure.

Never hire just one salesperson.

You need to be willing to commit to going out on the first 10 calls with new salespeople in order to make sure they are effectively communicating your company's message.

Are they performing the 11-act play exactly the way you wrote the script, or are they putting in their own additions that contradict it?

Ask a Vet

Ask any veteran sales manager what it feels like going out in the field with their people and they will all tell you the same thing, "I wish I could be out there every day."

It is absolutely amazing that, no matter how much you train people before they go out into the field, when they're out there on their own, they will say the wrong thing. You can't know that unless you are out there with them and hear them say it. Your correction—after the fact—will always be remembered by the salesperson who uttered the snafu.

If you plan on hiring salespeople and then retiring to your office to surf the Internet with your feet up on the desk, then the beginning of the end of your business has already started.

Training

Always commit half a day a week to in-office sales training.

This is the day when your salespeople are in the office, and you take the time to review every deal with them from the previous week. You spend time "role-playing" objections and rehearsing different acts from the play. You listen to them on the phone while they are talking to customers.

This is arguably the most important day of the week.

Talk to Your Salespeople Three Times a Day

Salespeople need lots and lots of contact.

As a sales manager, you need to know what is going on at all times. This isn't just because you don't trust them, but because you want to make sure their days are planned properly and that they are making the best of their time.

You want to be available at all times to celebrate the successes and to pick them up and brush them off when things go badly. Make three calls a day to your salespeople:

1. Start their workday with a phone call to review the day's plan.

2. Call around the middle of the day to check up on how things are going.

3. Make a final call, at the end of the day, to review the day's accomplishments.

This may sound like overkill, but the constant contact is vital. After 20 years in business, I still run a sales force this way.

Sales managers need to talk to their people three to five times a day. I still drop in unannounced occasionally to the morning and end-of-day calls just to make sure they are happening.

Salespeople need lots and lots of contact.

You want to be available at all times to celebrate the successes and to pick them up and brush them off when things go badly.

STILL WANT SALESPEOPLE?

If you are not prepared to do all of these things from the last few pages, then you really should not be spending the money to hire salespeople. Without following these steps (and there are a lot more!), your sales program will fail.

If you are, however, prepared to follow all the steps and understand the serious commitment that is required in setting up a sales team, then you have created your best opportunity for company growth and success.

EMPLOYEES—CAN'T GROW WITH THEM; CAN'T GROW WITHOUT THEM

You can't do all of the work yourself, right?

Yes, you can. You can do all the work yourself and you can do it better than anyone you could possibly hire. But is this really what you had in mind when you told your family and friends that you were going to go out on your own and start a business?

When you left your job and its accompanying paycheck, was it your goal to be doing the same thing you did before you left, only now with the headaches of being responsible for finding the work and managing the cash flow as well?

That wasn't your plan.

SO WHAT HAPPENED?

Every time I ask that question the answer is pretty much the same. Everyone starts telling me how hard it is to find good people who know how to do the work, or they launch into the story of how badly the last guy messed a job up.

Good people are hard to find. Most people available to work are not people that you want to have working for you.

Finding good employees is like finding a needle in a haystack. The players in our labor pool break down something like this:

Most people available to work are not people that you want to have working for you.

- **F players:** Not the least bit interested in having a job (5%).
- **D players:** Will work when they are broke and need money, but do a bad job (10%).
- **C players:** Barely willing to put in a day's work, but do so only because they need the money and the job (20%).
- **B players:** Will work when told exactly what to do and when followed up on hourly (30%).
- **A players:** Able to work independently and to solve problems as they arise (25%).
- **A+ players:** Able to run the business while you are on vacation (7%).
- **A++ players:** You (3%).

It is very hard for anyone who has ever had to do a lot of hiring to argue with this very unscientific study.

The challenge that most contractors run into when hiring help is their unrealistically high expectations of each applicant. They assume that everyone they hire will treat the job, the customer, and their tools with the same respect and dedication that they do. Unfortunately, that just isn't in everyone's genetic makeup. If it was, then everyone would have their own business and there would be no one left to hire.

> *The challenge that most contractors run into when hiring help is their unrealistically high expectations of each applicant.*

The Groups

The "F" Players—Hurricanes

We all know at least one of these guys. In some cases, we have them in our families or maybe they're a good friend who is great to go have a beer with (as long as you are buying).

Some people have no interest in doing anything productive. They just don't want to work. In extremely rare situations, when their backs are against the wall, they will work for short periods of time.

> *Some people have no interest in doing anything productive.*

These short periods of time are often referred to as "hurricanes," because of the rarity with which they show up on a job site, as well as the damage they leave behind. These natural phenomena rarely occur twice in the same location as these wonders of mother nature are never invited back to work again.

One of the most important things to remember about "F" players is that they are hired because they are generally charismatic folks who people enjoy being around. They have plenty of time to come up with stories to explain the large gaps in their employment history. They will infiltrate your organization and lay waste to a job site and customer relations if you don't cut them off quickly.

The "D" Players—Tropical Storms

These folks are generally employed regularly throughout their lives. They can easily be spotted by their general lack of caring and their colorful work history spanning multiple companies and industries.

Chances are you have employed or do employ one or two of these "D" players now. These are the guys who, no matter how many times you tell them to double check their work, always seem to miss something. They not only watch the clock, but make sure they start packing up for a break, lunch, or the end of the day a good 30 minutes early just to make sure that nothing cuts into their time.

Left to their own devices, they will mess something up eventually, maybe not always, but eventually. These are the guys who you yell at daily because you have tried everything else and nothing works. These are also the guys who you keep around because you gave up on finding anyone else.

"Tropical storms" do not do as much damage to your job site and business as a "hurricane," but they will generally leave a good-sized mess that needs to be cleaned up.

The "C" Players—Occasional Showers

The "C" players are those who most small contractors end up with. (We will get into the reasons for this in a couple of pages.) For the most part, these guys will get the job done, but only with maximum supervision. They don't disappoint as often as the "D" players do, although they do have a tendency to show up late from time to time and don't do what they say they are going to do.

The "C" player is the reason why most contractors have the feeling that there just aren't any good workers out there. They will be doing just fine for weeks on end and then, almost like clockwork, they will have a problem at home or with a customer that mushrooms into a major issue. At those times, you begin to question yourself as to why you started your company in the first place.

This group tends to be the biggest disappointment, because for many of us, this quality of employee is as good as we get. Your days will be going along just fine, when out of nowhere comes this rain cloud that drops crud all over your day. Maybe not enough to ruin your week, but surely enough to mess with your day.

The "B" Player—Sun Showers

These "B" players are often confused for "A" players, mainly because many of us have never actually had an "A" player working for us and therefore have nothing to compare the "B" players against.

The "B" player is hard working and gets the job done. They are conscientious and genuinely care about what you think of them and the job they do.

The only challenge with "B" players is that they need constant direction and don't think very well on their feet. When a problem arises, they call you and need specific direction to solve it.

"B" players, though, are much better than the "Cs" and "Ds," who either don't recognize the problem or don't care. "B" players are the largest percentage of today's workforce and, they are the reason why "management" exists. Management is not only there to fire the "Ds" and "Cs," but to make sure the "Bs" know exactly what to do every step of the day. This is how the majority of the work gets done.

Sun showers will pop out of nowhere, as a little problem now and again, but generally, everyone works through the issue without lasting effects.

The "A" Players—Rarely a Cloud in the Sky

These are the guys you dream of having at your side. An "A" player is the kind of person who does not freeze up in front of a customer. They are those who you can give a job to and almost forget about it. You only need to check in by phone and drop by to say "hi" and share a joke or two on the job site. These are the kind of people who will manage and motivate the "Bs," while isolating the "Cs" and firing the "Ds."

Stay tuned as to where to find these "mystery people" and, more importantly, how to keep them.

The "A+" Player—Sunny Days are Here Again (Your Vacation Players)

I like to call this group the "vacation players."

These are the ultimate people to have in your business. Many of you have never had one. Finding one is like trying to find a glimpse of a tiger in a dense jungle. They are elusive, but when you see one in its natural habitat, it is amazing to watch.

The "vacation players" are so named because these are the people who allow you to go on vacation and not worry about your business. The "A+" players are the ones who make all your problems go away. They treat your business as if it was their own, and they are the single most important group to have in your organization if you ever want to grow it.

The "A++" Player—The Weatherman

This is you. The risk-taker. The person who had the vision to start a business in the first place. You are the one who cares most about how things work out.

The key thing to remember about "A++" players is that if you are looking to hire people just like you, then you are going to be disappointed. They already have their own companies and are not looking for a job.

You create your own weather and show others how to navigate through the challenging storms.

FINDING PEOPLE

Now that we have identified the different players in the game, we need to figure out how to get them. The first thing to remember is that finding good people is the most important job you have as a business owner.

If you fail at this part of the job, you will never grow your business and you'll be stuck in a never-ending cycle of unhappy customers and long, frustrating days.

Scouting—Always Be Recruiting

"A" players are very difficult to find. They are generally not looking for work. They are employed by someone else. If their bosses are smart, they recognize who they have working for them and will do anything to make sure that they don't lose them. Lucky for all of us, most employers are not sharp enough to see the true value in an "A" player and, consequently, there will come a time when they are ready to change jobs.

"A" players can be found just about anywhere—not solely in your trade. In most trades, they can be taught and will learn the nuances of your particular trade. The skills an "A" player brings to the table are skills they are born with, not taught while on the job.

Constantly be on the lookout for people who amaze you. Look everywhere—the grocery store, fast-food places, or at other job sites. You need to think of yourself as a talent scout. Always carry business cards, your calendar, and a pen. Walk up to people and get their names and phone numbers. Ask if you can call them. Sit down and talk with them about a new opportunity.

Your Favorite Sports Teams Do It

Don't doubt yourself. Don't be tempted to tell yourself, "I can't do that. I can't go and hire away someone who is working for another company. It's just not right."

If that's the case, then prepare to be very frustrated and unable to grow your company. Just like a top sports team cannot get any better without constantly attracting top athletes, your company cannot grow without its "A" and "A+" players.

And don't count on attracting these "A" players through the want ads of the local newspaper. Why? They are already working for someone. You need to go find them!

The key thing to remember about "A++" players is that if you are looking to hire people just like you, then you are going to be disappointed.

You create your own weather and show others how to navigate through the challenging storms.

The first thing to remember is that finding good people is the most important job you have as a business owner.

Lucky for all of us, most employers are not sharp enough to see the true value in an "A" player and, consequently, there will come a time when they are ready to change jobs.

You need to think of yourself as a talent scout. Always carry business cards, your calendar, and a pen.

"No" Does Not Mean "No" (In This Case)

Persistence pays off. Most "A" players are not willing to leave their jobs on a whim. They are generally being treated well and don't see a need to leave.

The timing needs to be right. In many instances, it might take you months to get a great player to come join your team. It can be quite the task—phone calls, lunches, dinners, etc.

The time will come when the door opens for you if you've been persistent in recruiting the "A" players. Work slows down from time to time. Many of their employers, who don't recognize how lucky they are to have that "A" player, don't hang onto them during these slow times for one reason or another.

If you stay in touch with these "A" players, you can always count on a time when they will have something happen at work that will cause them to want to entertain other offers.

> *The time will come when the door opens for you if you've been persistent in recruiting the "A" players.*

Keep Your Roster

In my own business, I have always kept a list of people that I stay in touch with in order to be ready to have them join me when the timing is right. There will come a time when the right opportunity comes up for both of us, and the "A" player is ready to listen.

Patience is the key.

"DANGER, WILL ROBINSON!"

Don't be naïve. Your own "A" players are being stalked by your competition. People offer them jobs once every month or so. Sometimes they entertain these offers and sometimes they don't. Their decision to listen or not is 100% determined by how well you are treating them. Are they well compensated? Appreciated? Challenged? Do they see a long-term future in your organization? If not, then chances are they will be leaving you soon.

> *Don't be naïve. Your own "A" players are being stalked by your competition. People offer them jobs once every month or so.*

RECRUITING THE NORMAL WAY

Craigslist™

Craigslist is simple and it's free. It is a great way to gather a bunch of names of people who are sitting at home looking for work.

Keep in mind, though, that if there are 50 people applying for the job, there are only 10 who are probably worth talking to and only two or three who are worth hiring. Those two or three worth hiring may not want to work for you.

Newspapers

Depending on where you are located, newspapers can get you a long list of people who say they are willing to work. These ads can be expensive and, just like craigslist, require a ton of energy because only one in 50 people who apply are worthwhile.

Career Websites

Career sites are best suited when hiring for jobs that require a college education. They tend to be expensive and the results are similar to what you'd expect to receive from craigslist, newspapers, etc.

Friends of Friends

My favorite way to find people is to talk to friends or to people who are already working for you. They may know someone who is looking for a job in your trade.

My favorite way to find people is to talk to friends or to people who are already working for you.

If your employees enjoy working with you and feel like the company has a future, they are generally willing to refer you to their friends. Offering a hiring bonus of $200 or so is a nice incentive to have people start thinking about who they know. Make sure the bonus is only paid when someone has been with you for 90 days.

These referrals are prescreened. Your employees do not want to embarrass themselves by having you hire one of their lazy friends.

Now, if you are getting referrals from "D" and "F" players, all bets are off. Guess what caliber of worker they're going to recommend?

SCREENING APPLICANTS—HIRE SLOW, FIRE FAST

Hiring the first person who responds to your advertisement is generally a bad idea. With unemployment as high as it is right now, there are more people looking for work than there are good jobs.

You can, and should, take your time to go through the applications that you receive. Pick the best ones for further screening.

Top 15 Phone Interviews

Once you have your top 15 applicants, get your list of questions out and start calling the applicants for your initial prescreening. Your goal is to screen this list down to seven or eight people who you are willing to meet in person.

In-Person Interviews

Where?

In-person interviews can be held at your office or at any coffee shop in town. You just need to make sure that the environment is quiet enough and not too distracting to allow for a real conversation.

How Long?

Each interview will last anywhere between 30 minutes and two hours depending on the job.

A laborer may only need 30 minutes, while a foreman or other position of responsibility will take longer. Salespeople and other management roles will require multiple interviews by yourself and others in your organization.

Questions

All of your questions should be open-ended, meaning they should not be able to be answered with a simple "yes" or "no" response. Each question should require the applicant to have to think and explain themselves.

All of your questions should be open-ended, meaning they should not be able to be answered with a simple "yes" or "no" response.

Listen Carefully

Every answer that an applicant gives to you holds the clue as to what to ask next. If you listen closely enough, you will be able to find something about their answer that will lead you to another question. An interview is just like peeling the layers of an onion.

An interview is just like peeling the layers of an onion.

"'A' Players Are Immune to the Economy"

No matter how bad the economy is, "A" players will not be out on the street for long. They will always have options. You are not the only one who is hiring right now. Regardless of what you see on CNBC, people are still hiring, and the best of the best have options.

It Is a Two-way Street!

An interview is a two-way street. Why should this person come to work with *you*? What makes you and your organization special? Why do your customers buy from you?

An interview is a two-way street.

Part of the interview's purpose is to present potential employees with your company commercial. Once you feel comfortable that you are sitting in front of someone who you want to hire, be prepared to sell them on why they should join you. You want them leaving the interview with a great desire to join your company.

IT'S YOUR TIME—USE IT WISELY

Delegation is the key to a happy life.

Delegation can be broken down into two important categories:

1. **Things you should not do because others can do them better:** Recognizing your weaknesses is as important as understanding where your talents lie.

   For instance, I am not detail oriented; therefore, it is important that I surround myself with people who are great at details.

   I have tons of new ideas, just don't ask me to implement them without a team of people to clean up after me. I am not very good with money, so I have a top-notch CFO, controller, and accounting team (not to mention how I'd be a credit nightmare if my wife didn't manage our household bills).

   Spend the time to figure out what you are really great at. Then spend the time to figure out where your weaknesses are. Surround yourself with people who can fill in the gaps where you are lacking.

2. **Things that you should not do because it is better to hire someone else to do them:** We all know how this works.

   Should you be sweeping the floors on a job site when you can pay someone $10 an hour to do it?

STEPS TO FREEDOM

Every one of us knows the drill here, but there is a little trick to the whole thing. Here is how it all works.

First . . .

Figure out how much money you believe you should realistically make or want to make for a 12-month period.

Let's say the number is $100,000, maybe $250,000.

Second . . .

Take that number and divide it by 2000.

This will give you a very close approximation to the hourly wage that you are worth. $100,000 per year is roughly $50 an hour; $250,000 is roughly $125 an hour.

Third . . .

Here is the easy step. Hire people to do the jobs that are not up to your hourly wage. In other words, if the task would cost you $25 an hour, then you should be paying someone else to do the job.

All of this is simple, and this is obviously not the first time you have heard it. It's the fourth—and most important step of the process—that most people mess up.

If you don't do the fourth step perfectly, you will end up making a lot less money than the people you are employing for $10 an hour.

Fourth . . .

You need to make sure that your time is being spent doing the tasks that warrant your decided pay level. Tasks that are worth the bigger bucks are generally tasks that have to do with driving your business. Things like marketing your business, doing sales calls, managing multiple crews, and business development.

Tasks that will quickly have you making less than your employees would be things like surfing the Internet or napping in your truck and thinking about how you should really get out there and do something productive today.

Remember, in order for all of this to come together, you need to spend 40-plus hours a week on real tasks that drive your business.

> *Tasks that are worth the bigger bucks are generally tasks that have to do with driving your business.*

DISCIPLINE

Your business will fail if you do not have the discipline to spend your time on the larger priorities in your company. Simply hiring people so you can do less is a surefire recipe for failure.

As a business owner, you need to have the discipline that your employees may not have. You need to have the drive that others don't.

You need to be focused on what's important while others are playing or not paying attention. If you don't think you can do this on your own, then go get a job and save yourself a trip to bankruptcy court.

> *As a business owner, you need to have the discipline that your employees may not have. You need to have the drive that others don't.*

CHAPTER **9**

CUSTOMERS—YOU CAN'T LIVE WITH THEM . . .

For far too many home improvement contractors, the customer is the biggest obstacle that stops them from enjoying their favorite pastime—working with their hands and enjoying their trade.

Now, of course, not everyone looks at their customers this way, but let's be honest—we all have had those days or weeks where those pesky people get in our way.

EVERY CUSTOMER HAS AT LEAST ONE FRIEND

Too many home improvement contractors view the job they are on as just that, a job.

The most successful contractors understand that they have failed to fully satisfy a customer if that one job does not turn into two or three jobs in the same neighborhood.

Just drive down the street. Take a look at all those houses and think about how many home improvements they need. The list is huge. Half of the houses you see either need or will need your service within the next six months.

There are always things breaking, needing to be changed, or updated.

The "honey-do" list at most homes is much longer than "honey" has the time or energy to complete. All it takes for that homeowner to be your customer is the right time and the right place.

We have already covered the importance of referrals in our marketing mix, so let's talk about how you make sure you get them.

The "honey-do" list at most homes is much longer than "honey" has the time or energy to complete.

THE BASICS

There are two main reasons why customers are disappointed with their home improvement contractors:

1. Poor communication.
2. The mess on the job site.

More than any other reasons, these are why our industry has a bad reputation.

They are also the reasons why it is so easy to make money in this industry. The bar is set so low by those who don't understand the importance of these two simple issues that the rest of us are able to clean up, so to speak.

COMMUNICATION
Call Me Back—Please!

The most important reason why customers lose that loving feeling is because you don't call them back in a timely manner. The most important thing to remember here is that you respond *immediately* when a customer calls you, emails you, or texts you.

It is no longer acceptable to wait until the end of the day or end of the week to respond to a customer's inquiry. In today's day and age, our customers are used to getting the answers to their questions instantly. If Google™ can tell people how many penguins live on the South Pole in 2.6 seconds, then they feel you should be able to return their phone call, email, or text within 30 minutes.

Yes, You Do Need to be Able to Text

One of the keys to amazing customer relations is communicating in a format that makes your customer most comfortable.

Texting is not just for teenagers anymore. Texting has been around now for enough years that a bunch of those teenagers are now young homeowners. And many of the parents of those teenagers found it a great way to stay in touch with their children. Having discovered texting, adults now use it daily to communicate efficiently and effectively with friends, family, and coworkers.

Texting a customer who loves to communicate in this manner is like speaking the native language to those we meet on foreign vacations. It makes a very favorable impression.

Communication is vital. Find your customers' preferred method of communication and embrace it! Some prefer email, in-person daily chats, phone calls, texting, or a combination of the aforementioned.

Be ready and be aware.

Communication is Key

A key to a happy customer relationship is communication. The more you talk to your customers and the more you keep them informed of what you are doing, the happier they will be.

It isn't possible to over-communicate with your customers. If they ever feel you are making too much contact, they'll not be shy about telling you. But that just almost never happens. Feel free to email me and tell me that someone actually told you to stop keeping them informed about the progress of their home improvement project!

JOB SITE CONDITION
The Job Site—Hey, Someone Lives Here!

I am constantly amazed by the way some contractors forget that their job site is actually someone's home.

The most important reason why customers lose that loving feeling is because you don't call them back in a timely manner.

One of the keys to amazing customer relations is communicating in a format that makes your customer most comfortable.

Texting a customer who loves to communicate in this manner is like speaking the native language to those we meet on foreign vacations.

That home is also one of the most important things in that family's life. In fact, if you had to rank their most cherished things, the list would go something like this:

1. Kids
2. Spouse
3. House
4. Pets
5. TV
6. Car
7. Everything else

There are even some people who put their homes as number one, cars as number two, and . . . well, you get the idea. If a home improvement contractor misses this point, he will definitely not do a good job of getting their employees to treat the job site as someone's castle.

Off with Their Heads!

Everyone on your crews must understand that one of the top five "career-losing moves" is to disrespect your customer's home.

Everyone from you as the owner to the helper who sweeps up at the close of the day must *not* treat the job site as they would their own home (we all know why this is a bad strategy).

In keeping with the "castle" theme, they must treat it as if the king or queen of whatever country they respect most lives in that castle. If they disrespect the space in any manner, it will mean the gallows for them.

Everyone on your crews must understand that one of the top five "career-losing moves" is to disrespect your customer's home.

A Mess by Any Other Name . . .

A mess is a mess is a mess—all grounds for firing.

Steps for keeping the job site tidy:

1. All tools and supplies that are brought into or around a customer's home must be clean and well organized.
2. Drop-cloths should always be placed on the floor or ground, with all tools and supplies on top.
3. Paper booties need to be worn at all times in a customer's home.
4. All debris is to be cleaned up immediately, put in trash bags, and taken off-site. Do not use the customer's trash containers.
5. Never leave lunch or snack wrappers, pop cans, etc. at the job site.
6. Always perform a full cleanup at the end of every day. No exceptions.
7. Don't leave the little stuff until the end of the job—clean up as you go.
8. Always ask first about bathroom facilities, etc. Do not assume you can use them.

Job Site Cleanliness = Job Quality

Customers will always make the assumption that the cleaner, neater, and more organized the job looks during the work process, the better quality the job will be in the end.

When customers see a disorganized mess of a job site, they automatically assume that the quality of the job is in question and, therefore, they must "micromanage" the process themselves.

When this happens, your customers suddenly seem to be around a lot more than they were at the beginning of the job, they ask a lot more questions, and they seem to be asking for a job that was not the same one that you thought you signed up for. All of that can be avoided. Keep the place clean and organized!

Sharp-edged Tools and Other Things to Think About

Sometimes it is the simple things that get missed:

- Leaving tools lying around the job site where kids and pets can reach them.
- Chemicals left with the lid off for easy access to crew members and "Fluffy," the family cat.
- Ladders left leaning against a wall, for just a moment while you run out to the truck.
- All things that are tons of fun for kids and pets, but not for you after those kids and pets discover them.

When customers see a disorganized mess of a job site, they automatically assume that the quality of the job is in question and, therefore, they must "micromanage" the process themselves.

Magnifying Glasses

All home improvement customers own a huge magnifying glass and a set of binoculars that they choose to use or not use, depending on how you treat their home throughout the job.

The minute a customer feels that you are not to be trusted or that something is amiss in their home, they will go and grab their high-powered microscopes and turn them directly onto your job.

There isn't any home improvement job (or, for that matter, any other service or product) that can withstand that level of scrutiny.

If, by chance, your only crime is a messy job site, the next step is your customer questioning everything that you promised and did not promise.

All home improvement customers own a huge magnifying glass and a set of binoculars that they choose to use or not use, depending on how you treat their home throughout the job.

Lose, Lose

You will never win the battle of the messy job site. Your customers are right. They deserve a clean and neat home during the construction process.

You will either stand up to this level of scrutiny or end up paying for it in the end. You'll end up with an unhappy customer who will not give referrals and/or you will spend money out of your own pocket to make the customer happy.

Either way you lose. Remember—if your staff and crews do not understand this to be a top priority within your business, then "off with their heads!"

You will never win the battle of the messy job site. Your customers are right. They deserve a clean and neat home during the construction process.

We All Screw Up from Time to Time

So let's say that you and your crews agree that keeping the job site clean and organized is a top priority. And let's say that you have completed the last 10 jobs with no issues, but for some reason, this time you or your crew has checked their brains at the door and the job site is now a mess.

What do you do next?

Assess the Problem

Hopefully, you have found the problem before your customer has.

Either way, the first thing you need to do is to ascertain how big the problem is. Are we talking about forgetting to take out the trash? Did you use a different material than was requested? Did someone walk through the house with muddy shoes?

First, find out the scope of the problem and then immediately make a move to fix it.

You Can Run, But You Can't Hide

Remember, this is not your house. Your customer knows all of the best hiding spots and can tell immediately when something is out of place.

Like your mother always told you, take your medicine now or it will only get worse later!

Apologize Right Away

Don't wait until the customer has a chance to call you out on the mess you made. Most of them have kids and are tired of catching all the problems that are hidden from them daily.

Come right out and tell them what you found. Explain how sorry you are, and let them know what you are going to do to fix the problem.

The Biggest Missed Opportunity

Take the time to ask your customer if your solution works for them and ask them, "Is there anything else that I can do to make up for this mistake?"

In 85% of the cases, just the fact that you asked is enough to resolve their concerns and will put them at ease. In 10% of the cases, they will ask you for something simple, which is easy to provide. In 5% of the cases, there is no forgiveness. The customer really isn't seeking a solution. But, don't blame me. I told you not to screw this one up in the first place!

Hopefully, you will learn after getting burned once.

Take the time to ask your customer if your solution works for them and ask them, "Is there anything else that I can do to make up for this mistake?"

Icing on the Cake

At this point, 95% of the problems are solved. The customer is happy, and you are back at work. There is one final step you need to do to put this whole little mess behind you.

A simple, handwritten card and a $10 Starbucks® card saying once again, "Sorry for the inconvenience we have caused you, please enjoy a nice break on us."

Added Bonus

Feel free to use this on your personal life as well. I wish I did more of it!

TECHNOLOGY

Most contractors try to run and hide any time someone brings up technology. They're not interested in the latest gadget. They keep telling themselves the same cell phone they've had for the last six years works just fine. Hey, most of the buttons on that cell phone still work. (After all, do they really need to call someone whose phone number has a three in it?)

THE FIRST STEP

The first step in stepping into the 21st century is to stop being afraid of the new technology. You are not too old or set in your ways to learn how to use this technology.

We live in a day and age where customers expect to have their phone calls, emails, and even their text messages returned immediately. Customers simply don't understand why it would take you more than a couple of hours to call them back or return their email.

The first step in stepping into the 21st century is to stop being afraid of the new technology.

THE BASICS

Cell Phone

You need a good cell phone with all the basic options, and you need to know how to use them.

Voice Commands or Speed Dial

Keep the phone numbers of your office, key employees, and spouse in your speed dial numbers. Also, input your current customers into the speed dial numbers. Using the speed dial feature on your cell phone makes it much easier to call while on the road.

Using the speed dial feature on your cell phone makes it much easier to call while on the road.

Camera Phone

By having a cell phone with a built-in camera, you can easily snap a picture of everything from a problem area you want to show a customer to the finished project that you want to show a potential customer.

Even better would be doing a quick video of a finished project and sending it to a potential client's cell phone.

Even better would be doing a quick video of a finished project and sending it to a potential client's cell phone.

Text Messaging

Text messaging is not just for your teenage kids. It is a quick and easy way to communicate with your crews and customers.

Not all your customers text, but if they do, you'd better text as well!

Email

Email Address

A good email address will have the following qualities:

- Your name (first initial and last name or just first name, if possible).
- You don't want a series of numbers behind your name or something relating to your spare time like: fishingguy@. . . . com. You want to look like a professional company, not a weekday hobby.

Be @yourcompany.com

Gmail®, Yahoo®, AOL®, etc. are great providers of email, but do not give the image of someone running a real company. When your email address does not @yourcompany.com, you give the impression that you are running a small, disorganized, out-of-the-back-of-your-car outfit. This is not the image you are trying to achieve (not if you are reading this book).

Easy to Remember

The easier to remember and the shorter your email address is, the better. Ideally, you should be able to tell someone your email address on the phone, and they should have no trouble writing it down or remembering it.

Returning Emails

Consider purchasing, if you haven't already, a BlackBerry®/iPhone® or equivalent PDA. These phones allow you to read and reply to emails as easily as if you were at your desktop computer.

People expect that, just like a phone call or text message, their emails will be returned immediately, even if it is just a quick note saying, "I will respond in later detail tomorrow."

Choose your poison, but mobile access to email is just as important as a cell phone.

Signature Line

A signature line will have:

1. Your name
2. Your company name
3. Your title
4. Your cell phone, office phone, and fax numbers
5. Your company's tag line

Ideally, you should be able to tell someone your email address on the phone, and they should have no trouble writing it down or remembering it.

Choose your poison, but mobile access to email is just as important as a cell phone.

Spell Check

If you are like me, you should make sure you spell check every email before you send it.

Fax and eFax

Faxes are still a part of our daily business lives, and eFax makes it easier to capture and save every fax you get. No more lost faxes, etc. eFax will send each fax to your email address and save them until you can print them out, if and when you really need them.

BUSINESS INTELLIGENCE

There have been many books written on "business intelligence." For the purposes of this book, we will boil it down to the process of collecting information about the marketplace in which you operate.

In order to survive and thrive in today's business climate, you must truly understand the lay of the land, including all of your competitors, big and small.

THE MARKETPLACE

What is the geographic area in which you operate?

Within that area where are the real "prime" areas where you get, or should be getting, the majority of your work?

What makes this a prime area?

How many homes are in this area?

What is the age of the homes?

How old should homes be to benefit from your particular services, or does home age not matter?

What is the average household income in your area?

What are the general economic conditions in your area? Is it just average, unaffected by the economy, or are you working in an economic "ground zero" situation such as Detroit?

What are people in your area willing to spend money on right now?

YOUR COMPETITION

Understanding your competition can be a key part of your success. The more you know about your competitors, the better chances you have of making the proper decisions when a competitive situation arises.

Who is Your Competition?

List each of your competitors and answer the following questions for each of them:

- Name?
- How long have they been in business?
- In what do they specialize?

- How much business do they do in an average year?
- What is their reputation in the marketplace?
- Are they priced higher or lower than you? Why?
- What does their Website look like?
- What suppliers do they use and why?
- Are they getting the same or better prices on materials as you?
- Are they trying to grow, maintain, or panic and shrink?
- Who are their key employees?
- Are their key people happy?
- Do they have any recurring clients?
- Who are their best clients?
- What is their reputation on the Internet?
- What else do you know?

Now What?

Create a file on each of your local competitors and update it regularly. This information can help you in many ways.

Each of your competitors is better than you at doing something. Your goal is to figure out what it is and find a way to make your business more successful.

Knowing how your competition generally bids against you can be very helpful in a competitive situation. Understanding that they book jobs because they are pushing a new type of material or product that you had not thought about carrying could help you grow your business.

Knowing that a competitor's key employee has just left because they were unhappy could be a great way to pick up a talented person.

What if you found out that your competition's owner was getting ready to retire, and that you could pick up his or her clients for a nominal percentage for a few years of the business those clients produce each year?

Create a file on each of your local competitors and update it regularly. This information can help you in many ways.

How to Gather This Information

Ask questions and listen carefully. The more questions you ask of customers, suppliers, and your own employees, the more information you will obtain.

Your own employees probably know almost everything you need to know about all of your competitors.

As the boss, you should never:

- Spy on your competitors.
- Rifle through your competitor's trash.
- Create false reviews on the Internet about them.
- Outright copy their materials.

Your own employees probably know almost everything you need to know about all of your competitors.

- Ask them to bid your mother's house.
- Break any laws at all!

You can gather all the information you need, just by asking questions.
Things to remember about your competition:

- Bigger does not mean better.
- Years of experience does not always equal quality.
- Most employees are unhappy and think about leaving once or twice a year.
- Your competitors are just like you and will mess up a job from time to time.
- When they are full up with business, they will bid with a higher price.
- When they are slow, they will bid lower.
- We are all better in our niches.

BUSINESS INTELLIGENCE, OUTSIDE YOUR MARKET

If you want to be the biggest and the best, then you need to study who the biggest and the best are. The biggest and best may not be in your area, but you still want to know who they are.

Who are the premier players in your trade? Who has built the biggest, the highest quality, and the most profitable business in your industry?

If you want to be the biggest and the best, then you need to study who the biggest and the best are.

What Makes Them So Successful?

Create a file of the top five players in your industry. Answer the same list of questions for each of these large players. After creating the file on each one, keep filling it up. Bookmark the Websites of the top five players and review them Websites and review them. Ask suppliers to fill you in on what they know.

Use trade shows to gather information. In many cases, these large companies will be happy to share ideas, especially if you are not direct competitors.

In many cases, these large companies will be happy to share ideas, especially if you are not direct competitors.

A COUPLE OF FINAL NOTES ON COMPETITION
Friendly Is Always Better

It is a small world, and it is generally better to be on friendly terms with your competition. That is not to say you need to go grab a beer after work on Fridays with them.

You never know when you will come across a job that is outside your scope of work, and you may be able to hand it off to someone better suited to handle it. Do this and your competitors will return the favor.

Remember

Knocking or generally speaking ill of your competition to a customer will almost always lose you the job.

Knocking or generally speaking ill of your competition to a customer will almost always lose you the job.

BUSINESS INTELLIGENCE—STRATEGIC RELATIONSHIPS

Business intelligence is not just about your market and your direct competition. It is also about understanding who are the other trades in your area that market to the same customers.

Who are the top five . . .

- Painters
- Plumbers
- Roofers
- Handymen
- Landscapers
- Carpet/flooring installers
- Solar
- Pool people
- Remodelers
- Carpet cleaners

Create a file for each trade, and make it your mission to meet with the top few from each category.

Create a file for each trade, and make it your mission to meet with the top few from each category.

Why?

We are all in the home improvement business.

We are all marketing to the exact same clients.

We all do unique things that are better than what others can do.

If you are able to share and put into practice each other's ideas, you will only make each other's businesses stronger.

Alliances

Having an alliance with another trade will create more business for the two of you. When you come across a lead for another trade and share it with your new "business partner," you will have resolved a dilemma for your customer as well as ensuring that this "partner" will do the same for you.

If you were to create alliances with at least one company in 10 different trades, it would be like having 10 more connected people looking for business for you.

If you were to create alliances with at least one company in 10 different trades, it would be like having 10 more connected people looking for business for you.

The great thing is that just like many ideas in this book, as simple an idea as it sounds, most of your competition is not doing it!

BACK TO SCHOOL

In writing a chapter about going back to school, I'm not suggesting that you turn around and enroll in college, major in psychology, and start a new profession. I am simply suggesting that you make learning a priority in your life. Become an expert in everything important to your field. There are always new products, new systems, and new ways to do just about everything.

THE ENEMY OF LEARNING IS KNOWING

One of the biggest obstacles to growing a business is becoming absolutely convinced, because you have been in your profession for 15 to 20 years, that your way is the best and only way to do things. This leads to dangerous behavior.

The "Boss" Attitude

This mind-set can be deadly. If the owner of any business has it, creativity in the organization is stifled, and the employees feel like their opinions are not as important as the boss'. Over time, this kind of thinking will KILL your business. No single person has the "keys to the city" or the answers to all questions.

The Salty Dog

The salty dog is defined as the employee who has been in the business for years and years, typically having forgotten more about the business than you will ever know—and they are not afraid to tell you!

They are a huge value to the business because of all the knowledge that they bring, but they can also be extremely dangerous because they may believe there is nothing left for them to learn.

They are the ones responsible for the coining of the phrase, "The enemy of learning is knowing." No matter what you say, they will look you straight in the eye and say, "I already know the best way to do this."

Taking the Salt Out of Your Dog

There are a few things that you can do in order to have a good working relationship with this type of employee:

- **They need to feel respected:** Take the time to make sure they understand that you appreciate their opinions and respect their experience.

- **It is important that they feel like they're being heard:** This takes time and lots and lots of actual conversations. You are competing with years and years of history and experience in that noggin of theirs.

- **Ask, don't tell:** The best way to get them to try new things is to ask them to "test" a new method of yours and for them to determine which way would work best. *Telling* salty dogs to do anything will only get you barked at and possibly bitten!

> *The best way to get them to try new things is to ask them to "test" a new method of yours and for them to determine which way would work best.*

Not Invented Here

The organizational equivalent of the salty dog is the "not invented here" attitude. This mind-set generally takes over an entire company when its employees have been working together for an extended length of time. Every company has some degree of this within its organization.

The trick is not to let it infest too deeply into your workforce. The "not invented here" disease is one that tends to make all new employees feel like outsiders and is a compelling reason why some companies have a hard time attracting and retaining new talent.

HOW TO LEARN

Trade Associations—Get Involved

Every industry has its own trade association, and every trade association has local chapters.

Local Meetings

Local meetings are a great way to meet people in your industry who are doing things differently than you.

These meetings are almost always very friendly gatherings where contractors will get together once every month or so to listen to a guest speaker and have dinner.

These meetings give you a chance to share war stories with colleagues who may have already experienced your problems and may have suggestions for you to consider to resolve yours.

Conventions

Conventions are a great way to take a few days away from working "in" your business and spend some time working "on" your business.

Most associations will have regional and national conventions that you can attend. These conventions will have multiple guest speakers who will address topics that are relevant to your industry. Most conventions also include a large exhibit hall filled with suppliers and other vendors that are relevant to your business.

> *Conventions are a great way to take a few days away from working "in" your business and spend some time working "on" your business.*

Build Your Advisory Team

You don't know what you don't know! Very few, if any, entrepreneurs have built a significant business single-handedly. No one is an expert at everything in a particular field.

One of the most important pieces of the successful company puzzle is the "advisory team" piece. This characteristic is shared by all the most successful companies.

Having the right team to turn to when important decisions need to be made is critical to the long-term health of any company. The following are the critical advisors who you will need on your "team" in order to grow your company.

Legal

A strong attorney will help you see the legal pitfalls of your pending decisions. (More on this in Chapter 13.)

CPA

A solid CPA will help you make smart financial decisions for you and your company.

One of the biggest reasons companies fail is that they run out of cash! A good CPA can see that dead end coming and may help you avert it.

Banker

A good banker is someone who is a partner in your business. He is someone who takes the time to understand what you and your business need to grow and stay healthy. A good banker is someone who you should feel comfortable talking to and who offers suggestions to you on how your company can become even stronger.

Bankers have a unique perspective in that they don't specialize in one particular industry. They have the benefit of telling you what other companies are doing to succeed.

IT Professional

Your IT professional is a "key" team member. As your company grows, you will be more and more reliant on technology.

Your IT person should be someone who is "sized right" for your business. In other words, someone who is used to dealing with the challenges of $50 million companies is not the right person to make sure your small business has a Website and email.

The most important part of picking an IT person (other than that they know what they are doing) is how quickly they respond to his or her clients. You should choose someone who is available when you need them.

You also want to make sure that the potential "team member" will explain what are sometimes complex technological terms in language that you and your team can understand and that he or she will be there to help you grow when you are ready for expansion.

Having the right team to turn to when important decisions need to be made is critical to the long-term health of any company.

A strong attorney will help you see the legal pitfalls of your pending decisions.

A good banker is someone who is a partner in your business.

Bankers have a unique perspective in that they don't specialize in one particular industry.

Your IT person should be someone who is "sized right" for your business.

Industry Experts and, Yes, Competitors

Utilize your trade associations to build relationships with competitors outside your area.

Many of your fellow contractors have spent years banging their heads against the wall as they learned to overcome problems that you are just now starting to experience in your business.

With rare exceptions, there is not a business problem out there that has not already been seen and solved multiple times. The solution to your current issue will come by simply talking to the right people who have already dealt with the same problem.

Alliances with competitors who are in different geographic areas are a great way to share ideas and grow your business.

"Serial Entrepreneur"

A valuable team member is someone who has been successful at more than one start-up. The reason you want this person on your team is simple. These people have a proven track record of accomplishing something more than once.

There are those out there who are running successful new businesses because they just got get lucky. The stars aligned properly, they had the right industry connections, or they hired the right people. That's not to say that someone who has been successful in business was just lucky; but if you really want someone good to get advice from, find someone who has been successful multiple times. Nobody gets lucky over and over again. They will have something to offer because of their vast experience in starting up businesses. Chances are they have also failed at ventures, so they understand the warning signs better than anyone else.

If you can find someone like this, regardless of the industry they come from, they will offer you a very fresh perspective as you begin your path to success.

Mentor and Coach

Everyone needs a coach. From celebrity athletes to the President of the United States, successful people surround themselves with top-notch advisors. One of the most critical people in that mix of advisors is a personal coach or mentor. This person should be more experienced in business than you are and should be willing to hold you accountable for the things that you say you will do.

Everyone needs a coach.

A coach is someone you meet with, either over the phone or in person, for an hour or so once a week or twice a month (depending on your needs). They will help you do what none of your employees can ever do—hold your feet to the fire to get your goals accomplished. They will make sure that what you say you are going to do, you actually do. They are your confidant, your mentor, and someone who, if you choose them wisely, will be an instrumental part of your future success.

Your coach ideally is someone who has "been there and done that" and has the battle scars to prove it. He should be someone who you look up to and are willing to listen to. You must be able to trust them completely and be comfortable sharing everything about your business with them. If you hide things or keep secrets, your coach will not be able to truly help you in your quest for success.

Your coach ideally is someone who has "been there and done that" and has the battle scars to prove it.

CEO GROUPS

For those of you who have reached critical mass, you should apply for membership in CEO organizations.

Here are a few that I have had personal success with:

Entrepreneur's Organization™—EO

Fueling the Entrepreneurial Engine

The Entrepreneurs' Organization (EO)—for entrepreneurs only—is a dynamic, global network of more than 7,000 business owners in 38 countries. Founded in 1987 by a group of young entrepreneurs, EO is the catalyst that enables entrepreneurs to learn and grow from each other, leading to greater business success and an enriched personal life.

EO's Vision: To build the world's most influential community of entrepreneurs.

EO's Mission: Engage leading entrepreneurs to learn and grow.

EO's Core Values:

- Boldly Go!—Bet on your own abilities
- Thirst for Learning—Be a student of opportunity
- Make a Mark—Leave a legacy
- Trust and Respect—Build a safe haven for learning and growth
- Cool—Create, seek out, and celebrate once-in-a-lifetime experiences

The Entrepreneurs' Organization also operates the Accelerator Program and the Global Student Entrepreneur Awards in partnership with Mercedes-Benz Financial.

EO's Facts-at-a-Glance

- Total sales of all members worldwide: More than US $101 billion
- Total members worldwide: More than 7,000
- Total number of workers that members employ worldwide: 924,000
- Total number of chapters worldwide: 113
- Number of countries represented: 38
- Average member age: 39
- Average member sales: US $14.4 million per year
- Average member employees: 131

Young Presidents' Organization™—YPO

YPO's Mission

YPO's core mission is to develop "Better Leaders Through Education and Idea Exchange."

The Power of Peers

No other leadership organization screens its applicants as closely as YPO, because true peer-based learning is the foundation of the YPO mission and experience. YPOers learn from their peers, they exchange ideas, ask for advice, and share best practices in an open and trusting way. CEOs have access to a lot of information, to smart subordinates, and perhaps a few too many "yes-men." YPO's strength is a membership of qualified peers eager to learn and interact and grow as leaders.

Vistage™

With more than 14,500 members, Vistage International is the world's foremost chief executive leadership organization, providing unparalleled access to new ideas and fresh thinking through monthly peer workshops, one-on-one business coaching, speaker presentations from hundreds of top industry experts, social networking and an extensive online content library of articles, best practices, podcasts and webinars.

This top executive network started from very humble beginnings over five decades ago. One October morning in 1957, a Wisconsin businessman named Robert Nourse met with four fellow chief executives in the office of the Milwaukee Valve Company to test a simple, yet revolutionary idea—share knowledge and experiences to help each other generate better results for their businesses.

Soon this group of businessmen was probing, asking questions and making suggestions. They challenged each other and worked together to solve issues and grow. At that moment, TEC (The Executive Committee) was born.

My Experience

I have been a member of EO, YPO, and Vistage at different points in my career.

I have never been a good student. In fact, I am a college dropout with a GPA that my more educated business partners continually laugh about.

However, what I did realize when I started my business was that I needed education very specific to running an organization.

I have spent more than 14 years in a combination of EO, Vistage, and YPO and I am eternally grateful to the organizations, as well as to the friends who I have in each. They all share in the success that I have achieved.

As I look back on critical moments in my career, the people in these organizations were always there to help me navigate the choppy waters.

OTHER ORGANIZATIONS

While I have no personal experience with these organizations, I want to share other options with you.

Business Roundtable™

Business Roundtable is an association of chief executive officers of leading U.S. companies that has more than $5 trillion in annual revenues and nearly 10 million employees. Member companies comprise nearly a third of the total value of the U.S. stock markets and pay nearly half of all corporate income taxes paid to the federal government. Annually, they return $133 billion in dividends to shareholders and the economy.

Business Roundtable companies give more than $7 billion a year in combined charitable contributions, representing nearly 60 percent of total corporate giving. They are technology innovation leaders, with $70 billion in annual research and development spending—more than a third of the total private R&D spending in the United States.

Business Roundtable unites these top CEOs, amplifying their diverse business perspectives and voices on solutions to some of the world's most difficult challenges. Combining those insights with policy know-how, Business Roundtable innovates and advocates to help expand economic opportunity for all Americans.

Business Roundtable believes the basic interests of business closely parallel the interests of American workers, who are directly linked to companies as consumers, employees, shareholders, and suppliers. In their roles as CEOs, Business Roundtable members are responsible for the jobs, products, services and benefits that affect the economic well-being of all Americans.

Robust participation by member CEOs is a key strength of Business Roundtable. The organization is selective in the issues it addresses, a principal criterion being their potential impact on the economic well-being of the nation.

BNI™—Business Networking

BNI is the largest business networking organization in the world. It offers members the opportunity to share ideas, contacts, and most importantly, business referrals.

Using its "Find a Chapter" feature, potential members can contact a local BNI Director to see why BNI has brought businesses together all across the globe for over 24 years!

PERSONAL GROWTH—A COMMITMENT TO READING

Some of you love reading, and some of you are like me . . .

Those of you who are prolific readers, well good for you! You can officially skip this section (read on anyway).

Business Books

Reading fiction or the sports pages does not count. Growth comes from reading those dreaded business books—books that will cause you to rethink the way you are running your company.

To start, take just 30 minutes a day and read one book a quarter.

Trade Magazines

Subscribe to your trade magazine. These magazines will have tons of tips and ideas for running your business.

Subscribe to your trade magazine. These magazines will have tons of tips and ideas for running your business.

Websites

The Web has tons of sites that are dedicated to contractors. They have articles, forums, and blogs, and the writers come from around the country. This is a great way to stay informed about your trade.

EXPENSE MANAGEMENT— RUNNING A LEAN BUSINESS

Keeping an eye (or both eyes for that matter) on the bottom line has never been more important. In order to survive the economic storm we are in, we all need to pay attention to every penny being spent. No amount is too small to pay attention to.

MATERIALS

Materials Estimating

This is a simple but many times overlooked area of cost savings.

Knowing exactly how much material is required for each job is critical to being profitable. Are your estimating standards tight? Do you track the difference between the quantity of materials on which you bid and the actual materials used? Are there areas where you consistently underbid materials from job to job?

Material Suppliers

Have you explored all options with your suppliers?

Are you aware of all the products on the market today and how they may be different from the ones you are using? Are there products that could be saving you time and money if you used them instead of what you are currently using?

Are you spreading out where you make your purchases across too many sources? Can one supplier offer you better pricing, terms, or rebates if you buy exclusively from them?

Are you taking advantage of sales and bulk purchases on items that you know you are going to be using on a regular basis?

Inventory of Materials

Stuff walks off jobsites almost daily. In many cases excellent people who would never be caught stealing even a candy bar from the corner market will somehow justify taking home extra or unused materials from your job site.

Have you explored all options with your suppliers?

In many cases excellent people who would never be caught stealing even a candy bar from the corner market will somehow justify taking home extra or unused materials from your job site.

In many cases, workers feel that carting off with your materials is just something that "everybody does" and, therefore, they don't think of it as theft. Bottom line? If they take what isn't theirs without your permission, they're stealing from you and, as a result, are hurting your profitability.

What are you doing to control the consumption of your materials?

Controlling Inventory

What gets put on the plate gets eaten or taken. Develop a "just-in-time" method for supplying materials to your job sites. Everyone has seen how waste is always higher if a job has four or five days, worth of material stacked up than when crews are consistently worried about having enough materials to complete that day's tasks.

Take sandpaper, for example. (Even if you may not use sandpaper in your trade, your trade uses a consumable like sandpaper.) If you give a crew an entire ream of sandpaper at the beginning of a job, they'll only use somewhere between 10% and 50% of each sheet's potential. Check your trash every night and see what you find. If you limit the sandpaper to a couple of sheets per worker (depending on the job), you will find that each person will use 50% to 80% of each sheet.

It is simple human nature to waste things if you think you have an unlimited supply.

> *It is simple human nature to waste things if you think you have an unlimited supply.*

Get Angry!

Most employees have no concept of "your money." You are the "rich boss" and what does it hurt you if there is a little waste?

Visit your jobs and go through the trash. Is every roll of tape being completely used? Are they using too much? Are they getting every last bit of fertilizer out of the bag or getting lazy and leaving a cup or two at the bottom? Are they measuring twice and cutting once? Are they planning out a job to eliminate as much board waste as possible? Do they get the last drop of paint or glue from a can? Are they using five times as many nails as they need because they really like the sound of the nail gun?

You must instill in every worker's mind that you are watching everything on a job right down to the rags and nails.

> *You must instill in every worker's mind that you are watching everything on a job right down to the rags and nails.*

Focusing on Waste Does Not Mean Lowering Quality Standards

Paying attention to wasteful habits does not mean using lower-grade materials than what you should be using. Definitely do not substitute inferior products for what was originally included in the bid. Don't water down your paint or use two nails or screws where you should be using five. You get the point. Just pay very close attention to waste on your job sites.

LABOR WASTE—THE BIG ONE

How long should a task take? This is the most important question a home improvement industry business owner should be asking.

You need to know exactly how long every task on a job should take and, more important, you need to hold your employees to these standards. Your survival depends upon it. With a

> *With a proper estimating process you should be able to tell each and everyone on the job exactly how long each task will take to complete.*

EXPENSE MANAGEMENT—RUNNING A LEAN BUSINESS

proper estimating process you should be able to tell each and everyone on the job exactly how long each task will take to complete.

Holding them accountable is the next step.

How Do You Eat an Elephant?

One bite at a time.

Each job must be broken down daily into bite-size pieces.

How many feet of trench should be dug per hour?

How many windows need to be painted in an hour?

How many square feet of tile should be laid in an hour?

Expectations

Every worker on the job should know exactly what is expected of them by lunch and, then again, by the end of each day. This job falls on the foreman, superintendent, and/or yourself. It is absolutely critical to a job being done on schedule that everyone on the crew knows what is expected of them throughout the entire day—from the foreman, right down to the person pushing a broom.

The job will expand to fill the time available. If there is no clear-cut expectation of time set by you, then the tasks will always take longer than they should and the job will cost you more money!

Bonus Jobs for Coming in on Time

The only way to bring jobs in on budget is to bonus the people involved for accomplishing this goal. Put together a simple incentive system that pays your superintendent and foremen a small bonus on each job that comes in under your estimate. In other words, if you bid your job with a 40% gross profit and they are able to bring it in without sacrificing quality at 44%, then give them 2% of the overall job as a bonus. Everyone needs something to shoot for.

Everything Can Always Be Done Faster and More Efficiently

This must become one of your core beliefs.

There is room in every project to shave a point or two off of labor and a point or two off of materials just by being hyper-diligent and paying attention to every detail.

YOUR EQUIPMENT

How long should your equipment last? How long does the equipment last?

Do your crews treat the equipment as if it belonged to their old "shop" teacher? Do they even know where it is half of the time?

Do you have an inventory system and maintenance schedule?

If you answered "not sure" to any of these questions, you have some work to do.

Every worker on the job should know exactly what is expected of them by lunch and, then again, by the end of each day.

If there is no clear-cut expectation of time set by you, then the tasks will always take longer than they should and the job will cost you more money!

Put together a simple incentive system that pays your superintendent and foremen a small bonus on each job that comes in under your estimate.

There is room in every project to shave a point or two off of labor and a point or two off of materials just by being hyper-diligent and paying attention to every detail.

Vehicles

Many people don't treat company vehicles as if they owned them. They drive them too hard and don't keep up on the proper maintenance.

Put it in your schedule to spot-check your vehicles regularly. Expect your employees to treat the company truck like they would treat their father's vehicle back when they were scared of losing car privileges. Expect the vehicle to be clean and free of trash.

If you don't permit your employees to drive the vehicles on personal time, have you ever thought of installing GPS on your vehicles so that you know where the vehicle is being driven off the job? A GPS comes in handy, to check if your employees are out driving around town at 80 miles an hour or parked at a coffee shop for two hours.

Purchasing Power

Who is allowed to buy what in your organization?

What needs to be approved and what does not?

Is there a price limit on certain types of equipment that can be bought without your approval?

How do you track purchases to each job?

How closely do you review invoices for errors or extra items not needed for the job?

JOB COSTING

Every job that you do needs to be compared to the original estimate. It is imperative that the total amount of labor and materials for which the job was bid is compared to the actual labor and material cost on the job.

Every job that you do needs to be reviewed this way so that you can find areas for improvement. In some cases you may find that you are consistently underbidding a certain type of project or that your people are always taking longer in a particular area. Whatever it is, you'll find something that will save you money.

THE OFFICE

This is always a tough one. In some cases you need an office out of which to operate.

Others find a way to operate without the leased space. Even in our own businesses, I am constantly amazed by divisions that operate at the same revenue level despite some that utilize office space and the others that do not.

The ones that operate in a virtual environment are generally much more profitable.

The longer you are in a space, the more people tend to spread out. People move in and over the years they grab the cube next to them, the bigger office, etc.

Look at the space you have and ask yourself how much of it is really necessary?

- Is the space reasonable for the kind of work your company and employees are doing?
- Do you need the size of space that you have?

Many people don't treat company vehicles as if they owned them.

A GPS comes in handy, as well, to check if your employees are out driving around town at 80 miles an hour or parked at a coffee shop for two hours.

It is imperative that the total amount of labor and materials for which the job was bid is compared to the actual labor and material cost on the job.

- How often does the boardroom really get used?
- Could you operate with a storage unit and people working out of their homes?
- Do clients actually visit your space?
- When does your lease expire?
- Will your landlord give you a rent reduction to stay in the space?

Outsourcing

Outsourcing, as discussed here, is not about sending jobs to countries that most of us have never been to, but it is about finding a more efficient way to do the tasks that exist in your organization.

Some simple outsourcing examples are:

- **Payroll:** Payroll is a simple one to outsource, whether it is to Quickbooks™ payroll or ADP™. There is no reason for you to be cutting your own payroll checks and dealing with all the complications of tax withholding, etc.
- **Bookkeeping:** The simplest way to run your business is by using Quickbooks. I have heard all the arguments on other software, but, candidly, Quickbooks is what we use in our $47 million business and it works great right out of the box.

  Do you actually need a full-time bookkeeper, or can you outsource to someone who can do it part-time?

- **Phones:** Do you need to have your people answer the phone or can you find a professional service that will answer the phones for you and then text or email you the messages directly, giving you 24-hour coverage and saving staff costs?
- **Others:** What other roles exist that do not necessarily need to be fulfilled by full-time employees? What roles could you share with other companies?

  For example, we have had a Safety Consultant for years. For a monthly fee, he helps us negotiate all insurance coverage and makes sure that our safety programs are up to speed.

Do you actually need a full-time bookkeeper, or can you outsource to someone who can do it part-time?

Office Efficiency

Things can always be done faster. It does not matter what anyone tells you, take the time to watch a task and you will naturally see ways to accomplish it in less time. Owners are natural efficiency gurus. There is something about the fact that it is our money being spent that gives you an extra added sense of how to make things go faster . . . even when we have never done the task before.

Things can always be done faster.

THE LITTLE STUFF

The little stuff adds up quickly. Just take a look at your monthly financials, and add up all the below the line items. It is amazing how many costs are not attached to individual jobs.

Here are a few of those costs and some suggestions on how to bring them in line.

Cell Phones

Analyze your plans regularly. Are you getting the best price out there? Pay attention to the little stuff that adds up, like text messaging, roaming, ring tones, etc. Those "extras" make what you thought was a $100 a month charge become a $300 bill.

Offer employees $50 a month toward their bill instead of giving them a phone. They will take better care of the phone because it is theirs and they will be responsible for their own ring tones, etc.

Offer employees $50 a month toward their bill instead of giving them a phone.

Postage

Postage meters are expensive! In some cases they cost $100-plus per month. For years, postage meters were thought of as a necessary cost of running a business. Not anymore. Check out Stamps.com and get rid of your postage meter forever.

Miscellaneous Office Supplies

Are you paying full price at the local office supply store? Online stores offer a variety of solutions at amazing prices.

Phone Lines

What are you paying every month for your phone lines? How many lines do you have, and how many do you need?

Do you need a landline system, or could you change to an IP-based phone system? Imagine getting rid of all your landlines and using phones that are tied to the Internet instead. These IP-based phones have functions that most people never even thought were possible, all at a fraction of the cost of a landline.

What are you paying every month for your phone lines? How many lines do you have, and how many do you need?

Internet Service

Compare all of your options. Most of us are paying much more than we need to for our Internet service. You will be amazed at how much faster your Internet speed can be and how little you will pay for it.

Don't Buy New Furniture

Never, never, never buy new office furniture. There are plenty of used office furniture stores in every city in the country. You can also purchase your furniture on craigslist and other sites, just be careful not to spend too much time doing it.

I have never owned a new desk, and I never will.

Never, never, never buy new office furniture.

Cleaning Service

Do you have a cleaning service in your office? Shop around and you'll find significant differences in pricing for the same level of service.

Computers, Photocopiers, Etc.

Where do you buy your computers and equipment? Are you getting the best deal on equipment? Will it be fixed when you need it to be fixed, or are you just buying the cheapest stuff hoping it will work?

Price scrimping will hurt you here. Buying used computer equipment is not a good option.

I am generally not an extended warranty guy unless it is on expensive laptops and TVs. I have always gotten the better end of the deal buying the extended warranties on these items.

Check your leases on all your copiers to make sure you are getting the best deal for your money.

Where are you buying your ink cartridges? There are multiple options online to save you a bundle of money.

> *Price scrimping will hurt you here. Buying used computer equipment is not a good option.*

> *Where are you buying your ink cartridges? There are multiple options online to save you a bundle of money.*

Stuff You Don't Think About

There are other ways to save money in your office that may not be as obvious:

- Magazines you don't read.
- Bottle water service instead of a simple water filter.
- Electricity
  - Use CFL lights wherever you can.
  - Turn off computers and printers when not in use.
  - Use a power strip to avoid "phantom power" drain.
  - Set your AC at 72° and make sure it is off on weekends and nights.
  - Put your lights on motion switches and make sure they are turned off on weekends and nights.

THE BIG STUFF

Legal

Lawyers are said to be a necessary evil. It seems the longer you are in business and the larger you grow, the more interaction you have with this unique portion of the community.

Here are some tips for the next time you cross paths with this "necessary evil."

If You Are Being Sued

Check to see if you have insurance coverage for whatever the issue may be. You may be shocked at how many things you are actually covered for. You also may be shocked to see what you are not covered for.

Ignore the Dollar Amount

Most lawsuits have a huge dollar amount attached. This is done to get your attention and make you panic. Very rarely does anyone get anywhere close to what they are asking for (nor do they expect to).

Don't Panic

One of the purposes of a lawsuit it is to make you, your partners, and your family panic. The plaintiff hopes that you lose sleep at night so that you will be more motivated to write them a check quickly.

Yes, Some or All of What They Claim Is Untrue

Many lawyers will throw anything they can into the initial lawsuit. Claiming not only that you have done their client wrong (which may be true), they'll tack on how you have caused them to lose sleep and, of course, my favorite, caused them "sexual dysfunction."

Focus on what you know to be true and what you can prove. Don't worry about what they have made up.

Get to the Heart of the Problem Right Away

The first thing you need to do is to find out exactly why you are being sued. Get a clear understanding of what the plaintiff is claiming you did and what you actually did wrong, if anything.

Be objective. Very few lawsuits are just on the whim of a client or attorney. You did something wrong. What was it?

Focus on What You Know to Be True

Take the time to write everything out that you know about the incident in question. Write down every detail, including dates and times and anyone who can corroborate your side of the story. Have everyone involved write down all the details as well as they can remember them. Gather any documents, pictures, emails, etc., and get ready to tell your story.

Don't Hire the First Lawyer You Talk To

They say that it is a crime that you don't have to pass a test to become a parent (which I agree with). It is also a crime that all you have to do to become a lawyer is pass a test. Just because someone has passed the bar in your state and has a nice suit does not mean she is a good lawyer.

Get Referrals

Get at least three referrals from businesspeople you respect, not referrals from friends. Everyone knows a lawyer, but just because this recommended lawyer was great in a divorce case or DUI, that doesn't mean he is a good fit for you.

Make sure the referral is from someone who has actually used the attorney in the past and is speaking from experience, not just, "I heard she was a great lawyer."

What Do They Specialize In?

There are many different types of law, and they are all very different. Do not assume that just because a person may be a great real estate attorney that she knows how to defend you in a criminal trial (let's hope that never happens, but it is a polarizing comparison).

Check Out Their State Bar Record

Go online and check out the recommended attorney's record with the State Bar. Find out exactly what the attorney's history is and make sure he has a spotless record. While you are at it, check out the lawyer who is suing you. It is always beneficial to know your enemy.

Interview Three Lawyers

Lawyers worth their salt will give you an hour to see if there is a good match. Send the attorney a one- or two-page summary as well as the entire lawsuit.

There is never one simple way to defend yourself in a lawsuit. There are multiple approaches just like there are multiple lawyers. Find out how each attorney would approach the case. How will they charge you? How long do they think the process will take? Have they ever had a case like this before and, if they have, what was the outcome? Do they know the opposing lawyer?

You need to be 100% comfortable with the lawyer you choose.

Managing Your Lawyer—Speak Up, You Are in Charge!

If you don't like the way things are going, then speak up! You are the one paying the bills and if you are not happy with the direction that the case is going, then you have the right to say so. Lawyers do know the law. But it is amazing how many lack simple business sense.

Trust your gut, chances are you may be right!

Get the Case Closed As Fast As You Can

The most important thing to remember about getting sued is not to focus on whether you are right or wrong.

It is much more important for you to get the lawsuit to end as quickly and as cheaply as you can. This needs to be your and your attorney's focus. Good attorneys recognize that, if they get you out of your pain as soon as possible for as little as possible, you will then use them again and again.

You are in business. You will probably get sued more than once.

Getting your attorney to focus on closing the case quickly is the challenge. Both your attorney and the other side's attorney make their money by billing their time. Be very careful not to get caught up in long drawn-out letters, motions, discovery, etc. The world is littered with lawsuits where the legal bills were 10 times the actual settlement.

Review All Bills

You are the client and you have a right to review and question all bills. Lawyers bill for their time. Was it really worth it to spend three hours chatting about other issues or how your weekend was? They will charge you for that as well.

Being Right is Not Enough

Right and wrong have little to do with a lawsuit (even though we all wish they did). You don't have to search for very long to find the story of someone who was sued, was absolutely right, but still lost tons of money. Once again—if you are getting sued, your number one goal is to figure out a way to settle quickly for as little as possible.

Suing Someone Else

So you are mad. You have been wronged. You are totally convinced that it is absolutely clear that the other person is dead wrong and deserves to pay. You have suffered a tremendous amount of pain and suffering because of this wrongdoing. You are convinced that once a lawyer hears your claim, she will absolutely side with you and will work toward making everything right.

You *are* right about one thing—you'll be able to find a lawyer who will take your money and agree that you are right. Heck, some attorneys will fan the flames for you, and then get you all excited about how you have been wronged and how you deserve your day in court (all charged at their hourly rate).

Hold the presses.

If you want to see how they really feel about your case, ask them if they will take it on contingency. In other words, when they win this case for you, their only compensation is 30% of whatever is awarded to you.

They just sat there and told you that this was a slam dunk and that you were right, correct? Well then, they should be willing to do this for 30% of the victory, right?

Suddenly your new friend does not seem so excited, do they?

Instead, they tell you that taking cases on contingency is not their business model. They just don't take cases like that.

And why not?

Because attorneys know what you and I do not. Lawsuits take a very long time to argue, and even if you are 100% in the right, you can still lose. They know that at the end of a lawsuit, the two lawyers are generally the only two people leaving with their pockets full.

> *They know that at the end of a lawsuit, the two lawyers are generally the only two people leaving with their pockets full.*

Choosing Between Lawyers

If two lawyers are willing to take your case on contingency, pick the best and then see about just paying them hourly with a small success bonus. The only time a lawyer will take a case on contingency is when they truly believe that they will win no matter what and that they can make more money off of the 30% than off of an hourly fee. After all, they are in business to make a profit, too.

Contracts

Having a good contract is imperative to your business. But what is a good contract?

A good contract is one that is suggested by your trade organization as being "normal" for your business. If you want to understand the potential issues with your contracts and agreements in general, take any contract that you have had written by one attorney and take it to another attorney to review.

I have never once seen an attorney say "This is real quality work." They always say, "Well, this is okay, but I would change, this, this and this." Attorneys have a huge superiority complex.

> *Contracts, whether with a customer, vendor, or business partner, will occasionally end up in a dispute.*

Contract Disputes

Contracts, whether with a customer, vendor, or business partner, will occasionally end up in a dispute. A well-written and clear agreement is extremely helpful in all of these cases.

At the end of the day, though, these conflicts will generally end up in a negotiation that in many cases has little to do with how the contract was written and more to do with what was supposed to be the "spirit of the agreement."

In Summary Your Honor . . . Now What?

Years ago, I used to fight battles based on what I believed to be my sense of right and wrong. I took on all comers! No fight was too big or too small. If I believed that I was right and the other side was wrong, then I would take the fight all the way to the mat.

After many years of spending tens of thousands of dollars on legal fees and despite winning many cases, I forgot the most important part of the victory. There were many cases where I had won, but there was no way to collect any money, or the amount I won was nowhere near the amount I spent in legal fees or in collections. At the end of the day, I would have been better off just settling quickly or not suing and moving on with my business.

CPA Firms

A good certified public accountant will be an integral part of your team. They will be there to help you with overall business strategy. Think of them as a partner who you consult before going off in any crazy directions or spending large amounts of money. All entrepreneurs need someone to help reel them in from time to time. We get excited, and many of us, especially me, have never met an idea that we didn't like.

Running financial decisions by someone who is not as close to the day-to-day business operation, and who has seen others do similar things, can really give you a different opinion that may save you tens of thousands of dollars.

Equipment Purchases

Many contractors get caught up in the idea of having the best and newest gadgets.

Contracting is a very male-dominated sport, and most of us men have never met a "tool" we did not like. I still have this problem. As recently as last year, I thought, "Wow, with all the money we spend on boom lifts, we should just buy five and shuttle them around." Now, it's a good thing I have smart partners, advisors, a great CFO, and CPA firm, or I would have one of those things parked in my driveway most of the time (I would eventually get bored with checking out the peak on my roof) and still would need to rent different types of boom lifts anyway.

I am crazy enough that 14 years ago I actually thought of manufacturing my own paint. Fortunately, I once again had good advisors to steer me away from that one. Ironically, this came up again just a couple of years ago when a new partner said, "What about this idea?" I was able to help steer us away on my own that time by explaining to the new partner why this idea didn't make sense from a cost, manufacturing, or marketing standpoint.

Expansion

Expansion is a very expensive proposition, the costs of which are generally underestimated. A solid CPA will be able to help you put together the true cost of expanding your business and be able to look at your plan with a critical eye.

All entrepreneurs need someone to help reel them in from time to time.

Many contractors get caught up in the idea of having the best and newest gadgets.

I am crazy enough that 14 years ago I actually thought of manufacturing my own paint. Fortunately, I once again had good advisors to steer me away from that one.

A solid CPA will be able to help you put together the true cost of expanding your business and be able to look at your plan with a critical eye.

Acquisition

Before you even think that buying your competitor out is a great way to expand your business, have your CPA do a complete review of the other company's financials. Have your CPA tell you exactly where the downsides are. All you will see when you first consider the acquisition are the good parts of the deal and none of the downsides. Your CPA will show you the pitfalls.

Budgeting

Many home improvement contractors do not have a budget for the year. This basic but important part of business is something that your CPA can help put together for you. Based on your past history and your future expectations, your CPA can help put together a basic budget that you can review on a monthly basis to help make sure you are doing the right things to make a profit.

Job Costing

Good CPA firms will help you put together and/or review your job costing to make sure you understand what is working and what is not working on every job.

Interviewing Key Hires

Use your CPA firm to help hire bookkeepers, accounts payable clerks, accounting managers, controllers, and, if you are large enough, your CFOs. They will ask all the questions that you don't know to ask and will make sure you get the right person for the job.

Use your CPA firm to help hire bookkeepers, accounts payable clerks, accounting managers, controllers, and, if you are large enough, your CFOs.

Financial Controls

The amount of small businesses that have employee theft problems is staggering. In many cases, the culprit is a long-term and trusted employee who has been slowly and methodically skimming money for years. Your CPA firm can help make sure that you have the systems in place to safeguard your CASH!

Remember this phrase: "TRUST BUT VERIFY."

Remember this phrase: "TRUST BUT VERIFY."

CPA's Are Not Just for Taxes

You're getting the point. Many contractors only talk to their CPA firms when they are ready to file their taxes. If you bring your CPA in throughout the year to help check your progress and work on the overall business, he will save you much more money than you will spend on CPA fees.

No, You Are Not Supposed to Always Like Them

Having a CPA that thinks just like you and really likes your ideas is a dangerous relationship. The last thing you want or need in your business is a CPA who is just excited that you are paying them and agrees with everything you say. You need someone who will challenge your ideas. You want someone who will stand up to you and tell you from time to time that you are crazy.

If you bring your CPA in throughout the year to help check your progress and work on the overall business, he will save you much more money than you will spend on CPA fees.

Run, Don't Walk

Just like attorneys, the quality of CPAs is all over the map. There are many different business strategies in the world, and there are many different CPA firms using those different strategies. If you come across a CPA who is telling you about the newest "scheme" to avoid paying taxes or some special strategy that only a few of the smartest people are using, then I suggest you RUN, DON'T WALK down the street and interview another firm. If it sounds too good to be true, it generally is too good to be true.

Remember Arthur Andersen

Even big firms have bad ideas. Choose a conservative CPA and you will be protected and kept out of jail.

Choose Wisely

About 10 years ago, I felt it was necessary to change CPA firms. We had grown to a size that it made sense to move from a sole proprietorship to a larger firm that had a staff to be able to better handle our business. We were expanding into a good-sized, multistate operation and needed the resources of a larger firm.

Being the kind of guy who just wants to get things knocked off his list, I decide to move quickly. I found a firm recommended by a colleague. Since the firm was a large firm that had a specialization in contracting, I went ahead and had a meeting (which, of course, went well), and I signed them up on the spot. I was able to put a check mark next to "CPA firm" on my to do list, pat myself on the back, and move on to the next executive challenge on my list.

Later that week, I had my monthly meeting with my coach, who was also a Vistage chair. I sat down to lunch, went over the usual catching up on family and friends, and then got down to business. The first question was one I was very prepared for and excited to answer. "Where are you on finding a new CPA firm?"

I smiled and looked my coach right in the eye and said, "Done. Signed up . . . last week," and, of course, in a my normal cocky way gave him the, "You are not going to catch me this month" look.

Now, my coach is a very intelligent man who has been around the block many times. I was not the first or only eager entrepreneur who he had dealt with. He leaned back in his chair, smiled and said, "Great. How many did you interview?"

"How many did I interview? What do you mean? I got the job done, is that not what counts? If I found the right one the first time, why does it matter? Aren't all CPA firms the same? After all, accounting is an exact science. It's math. Isn't there only one way to do it right?"

I think I will remember that meeting for the rest of my business career as the one that really took the air out of my sails and reminded me why I believe every businessperson needs a coach.

It was not about getting the job done. It was about the process of learning from the meetings. Every CPA, lawyer, or other professional will approach problems from different ways. Those different approaches don't always match with your needs.

There is more than one way to skin a cat. After that meeting, I have always taken the time to learn the business version of the different cat-skinning techniques.

ESTIMATING AND PRICING

Estimating and pricing are two completely different pieces of your business.

Estimating is the "art" of figuring out exactly how long a particular task will take to complete.

Pricing is understanding your business model, your market, and your overhead well enough to be able to set a price for your services and turn a profit.

ESTIMATING SYSTEM

Having a system for estimating is a key component to your business. The "gut call" estimating system that many contractors use is just not going to carry you through the larger and more challenging jobs.

An estimating system can be a simple spreadsheet. On that spreadsheet is a breakdown of each task by crew hours and cost of materials. The system can also be a robust piece of software that will do it all for you.

Estimating Software and Books

There are some great estimating software programs, as well as books, on the market today. Research reviews on the Web to see what other contractors have to say.

THE HORRORS OF UNDERESTIMATING

Everyone has seen it and everyone has done it.

You quickly take a job that was supposed to be a simple project. You put some quick numbers together without paying much attention to the details.

Later you find out that it really took you or your guys twice as long as you thought because you took a guess at the scope of the job and didn't properly measure the project.

Here are a few reminders to help you avoid the horrors of underestimating:

1. **Pay close attention to the details:** Closely review each section of the job to make sure you fully understand the scope of work that will be required.

2. **Measure twice just to make sure:** Many a blown estimate is due to an improper measure.

Pricing is understanding your business model, your market, and your overhead well enough to be able to set a price for your services and turn a profit.

The "gut call" estimating system that many contractors use is just not going to carry you through the larger and more challenging jobs.

Many a blown estimate is due to an improper measure.

121

3. **Understand your labor costs and materials costs:**
   - What is the total average wage of the crew, including taxes and benefits?
   - What are the exact costs of all materials?
   - Don't forget the little stuff, making sure to include all the regular consumables that you will use on the job.

4. **Triple check your numbers:** Math errors are a terrible way to blow an estimate.

Math errors are a terrible way to blow an estimate.

PRICING

Once you have a clear understanding of what your costs will be for a job, you now need to mark up your job to account for your business overhead and, of course, your profit.

Your overhead should take into account all costs that are not directly related to the job. Include in this overhead costs associated with marketing, office, trucks, your salary, etc. These are all part of your overall overhead markup.

Markup on a home improvement job is going to be anywhere from 25% to 50%, depending on everything from overhead to market conditions.

Your overhead should take into account all costs that are not directly related to the job.

Understanding Your Markup

It is critical that you understand what you need to charge on your jobs in order to cover your expenses and turn the profit you desire.

Most of your competitors have the same basket of direct costs on a job that you do. For the most part, you are all pulling from the same labor pool and are paying roughly the same wages for your crews. Your materials cost should be very close to one another. The only difference from your price to your competitor's price is how much you each mark up the job.

Markup on a home improvement job is going to be anywhere from 25% to 50%, depending on everything from overhead to market conditions.

Prices Are All Over the Map

I am constantly amazed at how prices are so different when I get bids for work around my house. As a matter of fact, there are really only a few reasons why prices can be lower or higher.

1. The contractor either over- or underbid the job. This is generally due to a lack of an estimating system and/or not paying attention to the basic steps listed earlier in this chapter.
2. Contractor overhead is either very high or very low, and therefore the job is priced accordingly.
3. The contractor is using a cheaper or more expensive grade of materials.
4. The contractor is willing to take the job just to keep his guys working.

"Just to Keep My Guys Working"

This is a phrase uttered by too many contractors too many times.

They are way too willing to take a job that barely covers their costs just so they can keep their guys busy. This all-too-common practice only helps to lead to the contractor death spiral.

The concept of taking jobs just to keep your guys busy only leads to you having to spend hour after hour babysitting a job that is not making you a dime. The job never turns out the way you expected. The customer is never as cool as you thought he would be (considering the amazing discount they are getting and the fact that there you are, working for free).

Do yourself a huge favor and give your guys the time off to chase some side work so that you have the time to focus on what should be your number one priority—marketing and sales. It is impossible to focus on finding profitable work if you are spending all day on jobs that don't make you any money.

You Will Never Make It Up With Volume

Far too many people think that it is just fine to take work at little to no margins because they ultimately will make it up in volume. You are not a grocery store or a gas station. Ours is not a volume business.

If you were to book so much work that you increased your volume tremendously, your overhead would also increase proportionately. More work means more equipment, more trucks, more supervision, more office staff, etc. More volume always translates into more overhead. More overhead generally means less profit, not more.

JOB WATCH!

Understand the size and scope of jobs that you are comfortable taking. The death trap for many contractors is taking the job that is much larger than the jobs that are in their "sweet spot."

The Allure of Easy Money

A large residential or commercial job can seem like the path to riches. What's not to like? One job for a zillion dollars, all in one location, regular work for your crews, only one customer to look after, and *all that profit!* It's time to start looking at new boats!

If it was only that easy. Large jobs generally mean larger headaches. Not to say you should never go after larger jobs, but there are some things you need to think about first.

Type of Job

Remodel—Residential

Remodels generally mean a more aware and pickier customer. If you are used to working in the new construction industry, then remodels are a whole new world. Customers spend time looking over your shoulder.

The concept of taking jobs just to keep your guys busy only leads to you having to spend hour after hour babysitting a job that is not making you a dime.

It is impossible to focus on finding profitable work if you are spending all day on jobs that don't make you any money.

The death trap for many contractors is taking the job that is much larger than the jobs that are in their "sweet spot."

A large residential or commercial job can seem like the path to riches.

If you are used to working in the new construction industry, then remodels are a whole new world.

Bidding is sometimes from plans, but generally comes from a job walk-through with the customer. Sometimes there are "specs," but most of the time you are writing them up yourself.

Obstacles, from kids and furniture to existing landscaping, are now a part of your day-to-day operation that you and your crews may not be used to.

New Construction

This generally involves working for a general contractor (more about GCs later).

The flow of work is interrupted by the speed at which other subcontractors are doing their work. The inevitable issue of who is responsible for what task comes up regularly. You run into issues of what happens when another subcontractor damages the work you have already finished. Who is responsible?

Change orders are a significant part of how margins are saved and money is made. Bids are generally done as "take offs" from plans and not from a job walk-through.

Margins are generally tighter and payment can be slow.

You run into issues of what happens when another subcontractor damages the work you have already finished. Who is responsible?

Remodel—Commercial

Commercial work can involve retail centers or office parks that have completely different requirements than do other types of work.

For example, if you are working outside the building, what are the requirements for cars in the area? Do you need special permits because you are working next to the main street? Does the customer need you to perform the work at night as to not disturb the tenants?

Remodel—Industrial

Industrial work can involve an entirely different set of materials than you would use in residential or commercial jobs. Specifications are going to be drawn up, and your work will be reviewed by GCs and engineers. This kind of work is not for the inexperienced.

Industrial work can involve an entirely different set of materials than you would use in residential or commercial jobs.

Government Work

Government work is the new horizon. We all want our stimulus money!

Do your research here and don't rush into this type of work without fully understanding what is involved. Bonding requirements, etc. make this type of work tougher to acquire.

Government work is the new horizon. We all want our stimulus money!

Crews Crossing Over—Like a Fish Out of Water

Crews typically become very good at one specific type of work.

A crew that is excellent at remodel work will have a high degree of customer awareness and understanding of customer expectations. They will be used to dealing with all the ins and outs that come with this type of work.

Putting your crack residential crew into a new construction project is like putting fish out of water. Everything from the terminology to the flow of work is different.

The same is true when you take a great new construction crew and put them into a situation involving a remodel. All of a sudden expectations are different. Cleanup work that used to be performed by a completely different trade is now expected to be performed by them.

They are now expected to deal with the customer directly and, of course, all of their specific little requests! It is enough to drive a crew to quit.

Crews typically become very good at one specific type of work.

Government or industrial work is not impossible. You get the drift. I am not by any means saying that your crews cannot cross over. But you will need to understand the differences and be prepared to train your crews accordingly.

GENERAL CONTRACTORS

Doing work for general contractors is an art all to itself.

General contractors fall into three distinct categories: the good, the bad, and the ugly.

The Good

There are those general contractors who are amazing to work with and have a group of subs that they always use and pay regularly. These GCs understand the value of a good subcontracting network and, much like their own employees and family, go out of their way to keep subcontractor relationships strong and happy.

This doesn't mean you can perform poorly and still get work. But if you are a quality contractor who does what you say you'll do, you can create a long and strong business relationship with these types of GCs.

> *There are those general contractors who are amazing to work with and have a group of subs that they always use and pay regularly.*

The Bad

These general contractors are not capable of keeping their subs on schedule and, therefore, are constantly messing up your schedule. They are disorganized and poorly spec the work.

With these guys, your crews spend more time idle at job sites waiting for other subs to finish or redo their work. These GCs will fight you on every change order, and if you are one of the last trades in on a job, you will have to fight to get paid. Payment is slow, slow, slow.

> *Once they sign up the new subs, they grind them into the ground and hold them to the contract.*

The Ugly

These general contractors have zero respect for their sub network. They view subcontractors as a commodity.

They constantly pick subs with the lowest price and will keep getting bids until they find the one sub who they know unintentionally underbid the job. These GCs are confident that there are enough of you out there that someone will mess up and be low bid by a huge margin. Unfortunately, they are generally right.

Once they sign up the new subs, they grind them into the ground and hold them to the contract. In many cases this puts subs into a situation where they lose enough money on the job that they are forced to close down either in the middle of the job or after it ends.

The ugly GCs pay extremely slow on purpose and, often the final payment is negotiated at a cut rate because they know you need the cash to stay in business. These guys know you will take 20% less to avoid the risk of getting nothing.

> *The ugly GCs pay extremely slow on purpose and, often the final payment is negotiated at a cut rate because they know you need the cash to stay in business.*

How Do You Protect Yourself?

Do your research. Remember your business intelligence training and apply it here to all GCs for whom you are thinking of working.

> *Do your research. Remember your business intelligence training and apply it here to all GCs for whom you are thinking of working.*

Talk to suppliers, other subs, and anyone who knows anything about them to find out which of the three categories they fall into. Questions to ask:

1. How long have they been in business?
2. What was it like to work with them?
3. How do they pay?
4. How do they deal with change orders?
5. Do they know how to manage their sub network?
6. Are they organized?
7. Is their first job like this?
8. Were they low bidder and, therefore, will they need to grind all subs to make profit?

After getting the answers to these questions, you need to make your decision.

The carrot that the GC are hanging in front of you is big enough to launch you into the next phase of business, but if that carrot is being held out by The Bad or The Ugly, you *will* get burned!

OTHER THOUGHTS ON JOBS OUTSIDE YOUR AREA OF EXPERTISE

Cash Flow

Large jobs mean large commitments of cash.

Payroll will still need to be paid regardless of when you get paid for your jobs. Materials will need to be paid or set up on job accounts with joint checks so that you get the terms you need.

In many cases, large jobs don't pay you for 60 to 90 or even 120 days. If you are not set up to be able to fund the cost of not getting paid for 120 days, then you should not be taking these jobs.

Your business will not survive.

Large jobs mean large commitments of cash.

Bonding and Insurance

Large jobs may require specific types of insurance and bonding that you may not have. Read the job specs very carefully and make sure you are capable of providing the level of insurance, etc. that is required in the contract. There can be a huge price difference in carrying $1 million of insurance versus carrying $5 million.

Bonding on jobs is another requirement you cannot overlook.

Estimating

If you are not 100% confident in your estimating system (meaning being confident that you are dead-on accurate all of the time), then run away from this project.

Small errors on $10k jobs are challenging, but they become huge nightmares on $100k jobs. There is nothing like spending $125k to produce a job for which you collected $98k. This happens all the time.

Being Low Bid

If you find yourself with being the low bid on the job, try to find out where everyone else was.

If you find everyone else submitted bids within 5% of each other and you are 25% less, then you are about to lose your shorts!

FINAL THOUGHTS ON GOING OUTSIDE THE BOX

For the record, I am not trying to tell anyone to stay solely in their box and never try anything new.

What I am trying to say is that before you venture out into parts of the world that are new to you, make sure you do your homework!

Small errors on $10k jobs are challenging, but they become huge nightmares on $100k jobs.

For the record, I am not trying to tell anyone to stay solely in their box and never try anything new.

MANAGEMENT

There are so many schools of thought on managing people that when you search management books on Amazon, you get about 95,965 results. You could spend the rest of you career just reading management books and never actually manage a single person!

Since none of us have the time, unlimited wealth, or interest to spend that much time learning the nuances of every management philosophy, I thought I would distill my years of experience into a few simple paragraphs that have not yet failed me.

There are five basic aspects to managing people in any organization. These five components are the same, regardless of the industry or size of company you run:

> *There are five basic aspects to managing people in any organization.*

1. Job mapping
2. Understanding and communication
3. Expectations management
4. Incentive management
5. Culture

JOB MAPPING
The Right Seat on the Bus

The first part of managing people is to find people who fit the type of work you need to have done. You need to make sure the person you have in a particular spot has the skill set and attitude to actually get the work done. You can't take a person who is a salesperson by nature and try to make him an accountant. You can't take someone who is a computer programmer and send him up on a ladder and ask them to just hammer away (and don't look down). You can't take someone who really does not like dealing with the public and put him in charge of customer service.

> *The first part of managing people is to find people who fit the type of work you need to have done.*

One mistake we all make is taking those who are star performers with great attitudes and assuming that they will be good at anything we give them to do. That practice just does not work.

One mistake we all make is taking those who are star performers with great attitudes and assuming that they will be good at anything we give them to do.

Attitude Is Everything—Right?

A person's attitude is a key to having the right person in the right seat. But the question is, "What is the right attitude for the job at hand?"

It is easy to pick the excited person who has great customer service skills and who is always positive to be the person you want as your salesperson or dealing directly with customers. But do you really want that person as your bookkeeper or controller?

You want the person who is much more cautious, controlled, and just a little pessimistic to be in charge of your finances. You want someone who does not believe everything he hears from employees out in the field about when an invoice is going to be paid. You want someone who, in general, has a healthy distrust of those sales types. You want someone who feels the need to squirrel away nuts for a rainy day. In general, you want a bookkeeper or controller who will watch your back.

Skills Required for Each Job

Take a look at each job in your organization. What are the basic attitudes and skills required for each of the roles in your company? Make a list of the roles:

Sales _____

Office _____

Customer service _____

Superintendant _____

Foreman _____

Laborer _____

Other _____

What skills and attitudes are associated with each role?

1. _____
2. _____
3. _____
4. _____
5. _____
6. _____
7. _____
8. _____

Your Employees

Now make a list of your employees and gauge them on the same list of skills and attitudes. Do they match up with the role in the company where you have them? Is there a better seat on the bus for them?

Different Skills and Attitudes

If you do this simple survey on the roles and the individuals in your organization, you will see that you have people who are not suited to be in the roles they are in. In many cases, they are in roles that they thought you wanted them to have. They are not happy in them, but feel they would disappoint you if they told you that they wanted to change roles. This "job mapping" is the first step to having a functional and manageable company.

UNDERSTANDING AND COMMUNICATION

Many companies can be places of stress and frustration for the owners as well as the employees. Many people believe stress and frustration are just part of doing business in the contracting industry. They think working with homeowners and their expectations, deadlines, etc. make for the high stress level.

If you are experiencing stress and frustration in your company, it's not coming from the customers. The stress and frustration you are experiencing is coming almost entirely from the poor communication that is within your organization.

> *The stress and frustration you are experiencing is coming almost entirely from the poor communication that is within your organization.*

Communication—No, Not Text Messaging

This kind of poor communication is not about phones, emails, and text messages. It's the way in which we deliver the message and the way in which everyone hears the message we are delivering. We all communicate differently. For example, how many times do you think hard-driving entrepreneurs have frustrated their accountants for no other reason than the tone of the voice they're using with them?

The key to proper communication is to understand everyone's personality profiles. You'll know the best way to communicate your message to others if you know what makes them tick. They'll not only understand what you're trying to tell them, but will be excited to go out and do what you're asking them to do.

> *The key to proper communication is to understand everyone's personality profiles.*

Personality Profiles

Through a simple series of questions, personality profiles are able to give each person a 10-plus page description of who they are and, more important, how they deal with stressful situations.

They allow you to see just who the person really is. I have found them to be so dead-on accurate, that each time I start working with a new person, I send them a copy of mine. Armed with my personality profile, they know exactly what they are getting themselves into.

EXPECTATIONS MANAGEMENT

There have been countless studies conducted that do "deep dives" into why people fail at their jobs. The studies all come to the same conclusion. People fail at their jobs because they do not know what is expected of them.

Most people don't go to work each day with the intention of doing a really bad job that will possibly get them fired (of course you are right—there are exceptions to every rule and everybody has had one of these employees, but hold that thought for now). Most people go to work hoping to please their supervisors. The challenge, time and time again, though, comes when what you expect and what they *think* you expect are not aligned.

> *People fail at their jobs because they do not know what is expected of them.*

The Basics

Do your employees truly understand what your expectations are in the office and on the job site? If they were asked these basic questions, would their answers surprise you? For fun, take this short questionnaire and ask these questions to all of your people and see what answers you get. You may be surprised at what you discover.

> *Do your employees truly understand what your expectations are in the office and on the job site?*

1. What are the expectations for:
    -   Starting time?
    -   Breaks?
    -   End time?
    -   Personal and sick days?
    -   Vacations?

2. What do deadlines mean in your organization?
    -   Fast and hard.
    -   Soft and squishy.
    -   Do they matter at all?

3. Work product
    -   What are the quality expectations on your jobs?
    -   What should a work site look like from a cleanliness standpoint?
    -   Punch lists
        -   What is an acceptable punch list?
        -   How many items should be on a customer punch list?
        -   What does "done" mean?

4. Communication
    -   How long should it take to return a phone call?
    -   How long should it take to return an email?
    -   How long should it take to return a text?
    -   What is an acceptable message on voicemail?

- What should a signature line look like for your people?
- What does it mean to be on time?

5. Marketing
   - Does everyone have business cards?
   - Does everyone know how to take down information from a customer about another potential job?
   - Has everyone "role-played" and practiced how to talk to a customer about the company?
   - Do you give a bonus for new leads and does everyone know how much it is?

6. Appearance
   - What is your dress code?
   - b. What is your policy on hair grooming, jewelry, etc.?
   - What is expected for maintaining company vehicles?

7. Pay structure
   - Does everyone understand how and when they get paid?
   - Does everyone understand how they earn a bonus and when it gets paid?
   - Does everyone understand how they can earn a raise?

8. Advancement
   - Does everyone understand where they fit in the company and why they are important?
   - Does everyone understand what their next step would be if they perform above expectations?

9. Your company
   - Does everyone know your company's "commercial"?
   - Does everyone know the top reasons why customers choose you?
   - Does everyone know and carry with them your company references?
   - Do you practice these items with them regularly?

10. Consequences
    - Does everyone understand the consequences for not knowing the answers to the nine sections above?

Expectations Management—The Next Steps

It's Not the Estimate

Bringing jobs in profitably is all about how you manage a project. Estimating is plus or minus 10-20% at best. There are always unforeseen pieces of every job that an estimator is not able to predict. There are times when a bid is too low and you lose money. There are also times when your crew is cranking on all cylinders and a job gets done way under budget.

Yes, I know I just told you that having a great estimating system is a key part of your success. And it absolutely is—a *part* of your success, but not all of it.

Planning

A well-run job has a plan. The superintendent (this may be you) organizes the job and makes sure everything is ready to start on time. The customer is aware of the day and time, and everything is ready to go. There can be no crew downtime.

Materials

Materials have to be ready at the job site every morning. Downtime because of a lack of materials (which is bad planning) is the number two reason why jobs take longer than they are supposed to take.

Inventory

Foremen should be responsible for knowing exactly what materials will be needed on the job site for the next two days and for obtaining what is needed in advance of the next day's production. Foreman who can't do this should be trained or replaced with someone who can.

The Number One Reason Why Jobs Take Longer Than They Should . . .

The *number one* reason why jobs take longer than they should take is because the crew on the job is not clear as to how long things should take to accomplish. These crews have no expectations for how long tasks should take to accomplish.

How Long Does It Take?

Part of the "job planning" process is to break down all the tasks on a given job.

How long does each section of a given job take to finish? If you have a proper estimating system, then determining these important factors is an easy process of taking the estimate and translating it into a "job plan."

From the "job plan" you will be able to see exactly how many "man days" a job will take. From that number you can decide how many crew members will be needed on the job.

The Most Important Part Is . . .

Each day, everyone on the job needs to know exactly what is expected of them by lunch and by the end of the day.

This way they will know if the job is on schedule. You, as the owner, should be able to walk onto any job, and choose any crew member, and ask: "How far along on this particular task do you need to be by noon to keep this job on schedule?" If they cannot answer you, then this is why your jobs are taking longer than the time for which you've budgeted how much you budgeted.

INCENTIVE MANAGEMENT

Many of you believe you are paying people a great hourly wage and/or a great salary and that should be enough to get the job done correctly. You're right. You are paying to get the job done correctly. But is that all you really want? You are better served if your people go above and beyond just getting the job done "correctly" every day.

Foremen should be responsible for knowing exactly what materials will be needed on the job site for the next two days and for obtaining what is needed in advance of the next day's production.

The number one reason why jobs take longer than they should is because the crew on the job is not clear as to how long tasks should take to accomplish.

Part of the "job planning" process is to break down all the tasks on a given job.

You are better served if your people go above and beyond just getting the job done "correctly" every day.

What you really want is for your jobs to be brought in five points higher in profit than they are now. What you really want is for every customer not just to thank you, but to rave about you. If you really want higher profit margins and customers to be thrilled about you, then you need to bonus your people to do so.

Share in the Profits—Step One

Set up a program where superintendents and foremen share in the wins when they bring a job in at the quality you demand and with a profit margin that is higher than expected.

It is amazing the results you'll achieve when your supervisors understand that they will share in the win from having a job completed under budget. Your material waste will drop significantly as your foremen will now view the cost of the materials as coming out of their own pockets. They will quickly realize that any waste is counted against their bonus, and they will find new ways to make sure that everything is used properly. Miraculously, they will find new ways to getting a job done efficiently and what previously took three hours is now being accomplished in just two.

Inefficient workers will suddenly disappear from job sites as superintendents and foremen suddenly have no interest in keeping the slower people on the crew. At least one of your foremen will suddenly want to fire the same guy who you have been wanting to get rid of for years, because all of a sudden your foreman discovered how slowly he works. The difference now is that the foreman realizes that keeping his slow brother on the job is costing him money!

It is amazing the efficiencies that you will find on jobs when you tie your key people into the profit.

Share in the Profits—Step Two

Now that you have successfully implemented step one, you need to roll the program out to the rest of your workforce. Again, this does not have to require a massive dollar amount out of your pocket. Just incentivize enough so that everyone understands that, when a job comes in under "X" hours and "X" material cost, they, too, will share by receiving "Y" dollars. After implementing this step, you will have everyone in the field focused on getting your jobs done right and done fast.

Share in the Profits—Step Three

Office costs are one of those things that can creep up over time. Make sure that whoever runs your office receives a bonus to keep costs down and not incentivized to add people to your overhead. Work on a plan that gives the office staff a percentage of the money saved when the overall office costs come in under an agreed-upon amount.

Share in the Profits—Overall

When you put in the time and effort to get all of your key people to watch your bottom line (because they will share in the win), your life will suddenly change. No longer will you be the "lone ranger" who is the only one out there caring about the bottom line. You will have a group of partners who will care as much as you do about the company!

Your material waste will drop significantly as your foremen will now view the cost of the materials as coming out of their own pockets.

Inefficient workers will suddenly disappear from job sites as superintendents and foremen suddenly have no interest in keeping the slower people on the crew.

It is amazing the efficiencies that you will find on jobs when you tie your key people into the profit.

Make sure that whoever runs your office receives a bonus to keep costs down and not incentivized to add people to your overhead.

CULTURE—THE BINDING GLUE THAT HOLDS IT ALL TOGETHER

Your company's culture is decided by you, the CEO and owner. Every company has a culture. Every culture is as unique as the business itself. Employees wear their company's culture on their sleeves. It only takes a few minutes of talking to people within an organization to see what its culture is all about.

What Is Your Company's Culture?

Does it revolve around you and only you?
Are people afraid of what you will say or do?
Are your employees free to speak their minds?
Do you encourage the sharing of ideas?
Does everyone want the company to make money and succeed?
Do your employees socialize at work?
Do your employees socialize outside of work?
How does everyone define hard work?
How does everyone define play or having a good time as a group?
What do you want your company culture to be?

Write a few brief sentences that describe the ideal company culture for you:

_____

_____

_____

_____

_____

_____

Now What?

It is up to you to not only define your company's culture, but to make sure it exists. If people are excited to come to work every day, genuinely care about the well-being of the company and their fellow employees, and feel like they are trusted and respected, then amazing things will happen.

BUSINESS PLANNING

You need a business plan, and every plan should be written down. Here is an outline for writing a business plan straight from the U.S. Government's Small Business Administration Website at SBA.gov.

WRITING THE PLAN

What goes in a business plan? The body can be divided into four distinct sections:

1. Description of the business
2. Marketing
3. Finances
4. Management

Agendas should include an executive summary, supporting documents, and financial projections. Although there is no single formula for developing a business plan, some elements are common to all business plans. They are summarized in the following outline:

> *Although there is no single formula for developing a business plan, some elements are common to all business plans.*

ELEMENTS OF A BUSINESS PLAN

1. Cover sheet
2. Statement of purpose
3. Table of contents

I. The Business

A. Description of business
B. Marketing
C. Competition
D. Operating procedures
E. Personnel
F. Business insurance

II. Financial Data

A. Loan applications

B. Capital equipment and supply list

C. Balance sheet

D. Break-even analysis

E. Pro-forma income projections (profit & loss statements)

F. Three-year summary

G. Detail by month, first year

H. Detail by quarters, second and third years

I. Assumptions upon which projections were based

J. Pro-forma cash flow

III. Supporting Documents

A. Tax returns of principals for last three years

B. Personal financial statement (all banks have these forms)

C. For franchised businesses, a copy of franchise contract and all supporting documents provided by the franchisor

D. Copy of proposed lease or purchase agreement for building space

E. Copy of licenses and other legal documents

F. Copy of resumes of all principals

G. Copies of letters of intent from suppliers, etc.

UNDERSTANDING YOUR MARKET

You should have a solid written business plan that will force you to understand the market you are in and outline how you will grow your business.

THE CHALLENGE WITH BUSINESS PLANS

The challenge with creating long business plans is that while they are helpful in the beginning to get the ball rolling, they tend not to get updated as much as they should and sit on a shelf until the business either is highly successful or fails.

Many of you (including me) may not be big on writing a 35-page business plan, so here's a nice and simple five- to eight-page plan. This small plan tends to be used more frequently and is easier to update as things change.

Either way, both the large and small versions make great reading after the fact.

You should have a solid written business plan that will force you to understand the market you are in and outline how you will grow your business.

Many of you (including me) may not be big on writing a 35-page business plan, so here's a nice and simple five- to eight-page plan.

The Basic Plan You Need

Here is a basic framework for your plan. This is not the type of business plan you'd use to raise capital, but it is a plan in the practical sense. It includes things you need to know and understand about your business in order to be successful.

What business are you in?

- What is your business/trade?
- What types of work do you do?
    - Residential?
    - Commercial?
    - Industrial?
    - Remodel?
    - New Construction?
- What is your experience in each of these areas?
- Why did you choose to operate in these areas?

Where do you operate?

- What geographic location do you operate in?
- What geographic boundaries have you set, if any?
    - Why have you set these boundaries?
- What are the demographics for this area?
- All of this can be plugged in from the business intelligence section of this book.

Competition:

- This section is the complete business intelligence section that you put together earlier in the book.

Marketing:

- This section is everything you learned and will implement from the earlier branding/marketing/lead generation sections in Chapters 1, 2, and 4.

Financials:

- Your profit/loss projections for this year.
- Your profit/loss projections for the next two years.

Personnel:

- An organizational chart of the people you now employ.
- A projected organizational chart for 12, 24, and 36 months from now. This should match with your financial projections.

BUSINESS PLAN INCENTIVIZE SUMMARY

I truly do understand that creating these plans is not something that many of you want to spend your time doing. There are some people who are really good at writing fancy plans, and there are others who are really good at just working and getting the job done.

Regardless of which of these types of people you are, take the time and write down your plan. Use the resources you have in your company, your partners, employees, and board of advisors to help put this all together.

The financial plan and budget are probably the most important parts of the business plan and are something you'll want your CPA to help you with.

Review your plan (especially the budget) every month and make sure you are on track and that all things still make sense. Make adjustments as things change.

You will get much farther with a simple eight-page plan in which every item on it means something to you than you will get with no plan or with a 50-page plan you will never read again.

> *You will get much farther with a simple eight-page plan in which every item on it means something to you than you will get with no plan or with a 50-page plan you will never read again.*

I have spent over 20 years building my business. Without exception, at the conclusion of each year, I admit to myself, "I learned more this year than I did in any other year in business."

Even though I dropped out of college, I have dedicated a huge part of my adult life to learning and surrounding myself with people much smarter than myself. I realized early on that I did not have enough knowledge to be successful in business and that I needed to commit myself to my own education. I still feel the same way today as I did 20 years ago.

If you were to remember only one or two points from each chapter of this book, these would be my suggestions:

BRANDING

Once you have decided who you are and what you stand for (based on your customer feedback, of course), you and your employees need to walk the talk! Your brand is not just a catchy phrase; it is a way of life!

MARKETING

Marketing is a fickle beast. What works for one trade does not always work for another. Each and every market is different, and potential customers in those markets respond differently to advertising. In order to be successful in creating a sustainable lead program, you need to test each new marketing program and continually track the results.

WHEN THINGS GO VERY BAD

It is a new age of customer service; even the customers who are wrong are now right.

LEADS

A lead is just that, a lead. It is not a promise of a sale. It is up to you to turn each and every lead into a customer for life.

SALES, SALES, SALES

Customers buy from people they like and trust. Don't give them a reason not to like you.

SALES MANAGEMENT

There is no such thing as an independent salesperson. Every salesperson, no matter how amazing he is, needs support, training, and encouragement. If you are not committed to this process, then do not hire salespeople.

EMPLOYEES

Hire slow and fire fast. Understand that there is a big difference between recruiting and hiring. You want to be a recruiter.

IT'S YOUR TIME

The tools that got you to where you are today are not the tools that will get you where you want to be. Focus on your priorities as a business owner.

CUSTOMERS

The basics have not changed. You need to treat your customers the way that you would expect to be treated.

TECHNOLOGY

We will never go back to the way it used to be. You need to have mobile access to your email, and you need to learn how to text.

BUSINESS INTELLIGENCE

A CEO understands the market to which he is selling and the competitors who are in it.

BACK TO SCHOOL

You are never too old to learn. The way you historically have run businesses will never be successful again. It's time to learn new tricks.

EXPENSES

Every dollar you spend is a dollar less in your pocket for your family. Focus on the little stuff when it comes to expenses. If you don't, your existing profit will disappear.

ESTIMATING AND PRICING

Without utilizing a systemized estimating system, you cannot grow your business.

MANAGEMENT

Hold people accountable. Be fair, but don't forget to be fun.

BUSINESS PLANNING

The plan you drew up 90 days ago is no longer relevant or accurate. Review and revise your plan every three months.

This book is filled with a ton of information that I have learned over the years. My mode of learning came mostly the hard way.

If I did not go into as much depth as you would have liked on any particular topic, then I encourage you to build your advisory board, find a mentor, talk to your competitor, and get involved with your local trade associations.

You are the CEO of your company. You are responsible for everything that happens in your organization. The most important thing you can do for your business is to carve out time to work on it and not always in it.

Remember, if a simple guy like me—from a mining town in northern Ontario, Canada, who dropped out of college to paint houses—can build a business that spans multiple states and employs a couple thousand people, then being successful should be easy for you!